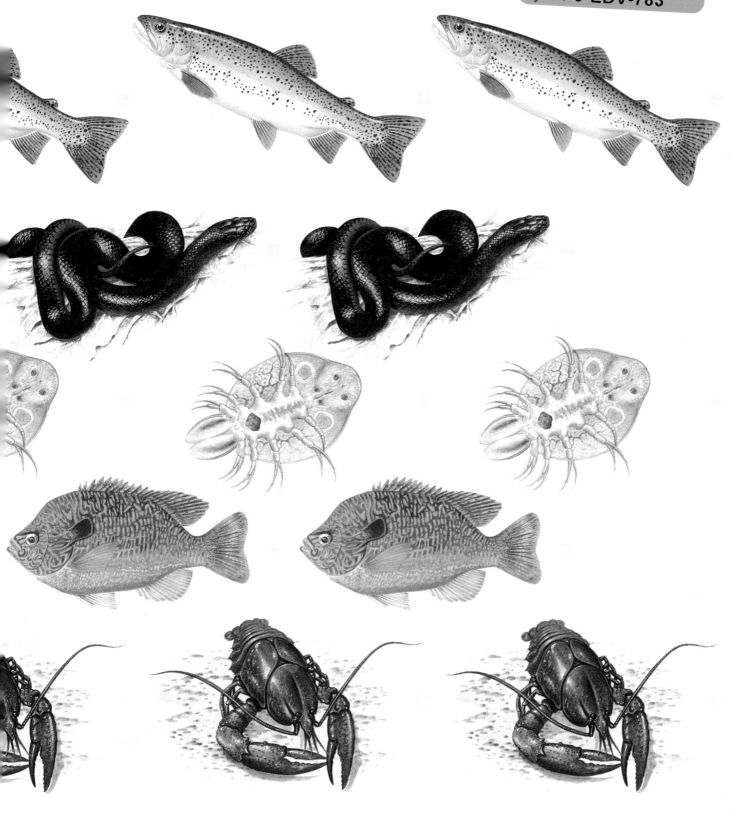

Poison Ivy and Poison Oak

This weed (shown above) causes an itchy rash if you touch it. Poison Ivy grows like a vine, and Poison Oak grows like a shrub. Try to remember what the leaves look like, and take care when walking in the country. If you do touch it, washing the area may reduce the itching. Your local drug store will have various remedies that will help.

Animal Sizes

One of these symbols is shown with each animal. It shows you at a glance how big the animal is likely to be. All measurements describe the length of the animal from the tip of its nose to the end of its tail.

Up to 1 inch   **1–2 ins**   **2–5 ins**

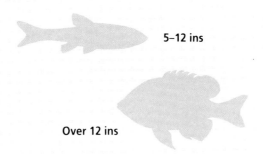

5–12 ins

Over 12 ins

For snakes, the measurements are shown by the symbols below, also from the tip of the snake's nose to the end of its tail.

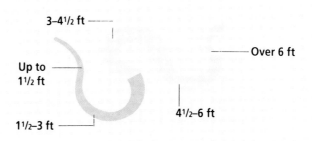

3–4½ ft ——

—— Over 6 ft

Up to 1½ ft ——

4½–6 ft

1½–3 ft ——

SCIENCE NATURE GUIDES

FRESHWATER LIFE
OF NORTH AMERICA

Susan McKeever

ILLUSTRATIONS BY
Colin Newman

CONSULTANT
Dr Frances Dipper

THUNDER BAY
P·R·E·S·S

Conservation

Every freshwater animal, from the alligator to the flatworm, is closely linked to its surroundings. It feeds on other animals or plants and makes its home in a pond, lake, fast-running stream, slow-running river, or in wetlands. Many animals prefer particular kinds of climate and landscape. As you learn about a habitat, you will get to know which plants and animals you can expect to find there.

Many of these habitats have been damaged or destroyed by industry and its pollution. People are draining wetlands—like flood plains, marshes and swamps—because they are good places to farm once the water has gone. Some freshwater animals are in danger of disappearing altogether—they are often protected by federal law, or by state laws.

On page 78, you will find the names of some organizations who campaign for the protection of particular animals and habitats. By joining them and supporting their efforts, you can help to preserve our animals and their habitats.

Countryside Code

1 **Always go collecting with a friend,** and always tell an adult where you have gone.
2 **Leave any wild animals that you find alone**—they may attack you if frightened.
3 **Leave their nests or dens untouched.**
4 **Keep your dog under control.**
5 **Ask permission** before exploring or crossing private property.
6 **Keep to footpaths** as much as possible.
7 **Leave fence gates as you find them.**
8 **Wear long pants, shoes and a long-sleeved shirt** in deer tick country.
9 **Ask your parents not to light fires** except in fireplaces in special picnic areas.

Thunder Bay Press
5880 Oberlin Drive
Suite 400
San Diego, CA 92121

First published in the United States
by Thunder Bay Press, 1995

© Dragon's World, 1995
© Text Dragon's World, 1995
© Species illustrations Colin Newman, 1995
© Other illustrations Dragon's World, 1995

Habitat paintings by Mike Saunders.
Headbands by Antonia Phillips.
Identification and activities illustrations by
Mr Gay Galsworthy.

Editor   Diana Briscoe
Designer   James Lawrence
Design Assistants   Karen Ferguson
                    Victoria Furbisher
Art Director   John Strange
Editorial Director Pippa Rubinstein

Complete Cataloging in Publication (CIP) is available through the Library of Congress. LC Card Number:

Printed in Italy

ISBN 1–57145–019–X

Contents

What Could I See?

From quiet ponds and peaceful lakes to rushing mountain streams and steaming swamps, freshwater areas are alive with animals. Water reptiles, like turtles, and amphibians, such as frogs, spend part of their time on land and part in the water. They have lungs and must come to the surface to breathe.

Some water snails have simple lungs and can live in still ponds where there is not much oxygen. Many insects start life in the water, then change into adults that live on land. But most other water animals, like fish, breathe using gills and will quickly die if taken out of the water.

This book will make it easier for you to identify all these water creatures in two ways. Firstly, it shows you only the animals that you are most likely to see. Secondly, it puts them in groups according to the habitat, or type of fresh water, where you are most likely to see them. But remember that many animals will be found in more than one habitat.

Body-changing insects

Some freshwater creatures undergo a change of form that is called metamorphosis.

Dragonflies begin their lives as eggs laid on a water plant by an adult. The eggs hatch into larvae called nymphs, which then live underwater for several years, hiding among the plants, and catching other animals for food.

Eventually, they crawl out of the water and change into adult dragonflies—they have wings, and can't live in the water.

This is called incomplete metamorphosis, because the young don't look that different to the adults. Stoneflies, mayflies, alderflies, and damselflies lead similar lives.

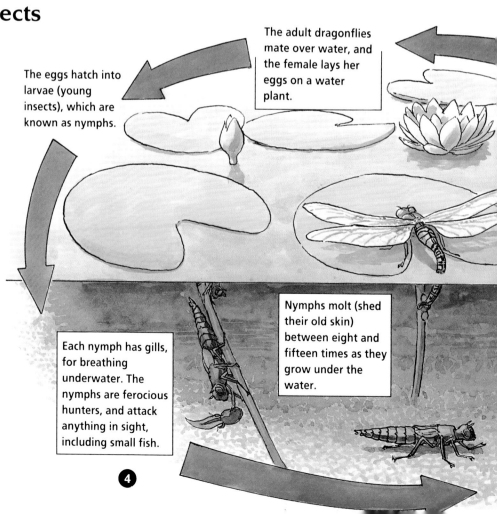

The eggs hatch into larvae (young insects), which are known as nymphs.

The adult dragonflies mate over water, and the female lays her eggs on a water plant.

Each nymph has gills, for breathing underwater. The nymphs are ferocious hunters, and attack anything in sight, including small fish.

Nymphs molt (shed their old skin) between eight and fifteen times as they grow under the water.

How to use this book

To identify an animal that you don't recognize, like the two animals shown here, follow these steps:

1 **Decide what habitat you are in.** If you're unsure about this, read the descriptions at the start of each section to see which one fits best. Each habitat has a different picture band heading and these are shown below.

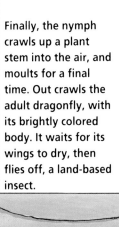

2 **Decide what sort of animal it is.** Is it a reptile or an amphibian, a fish or a mollusk, or something else entirely? Look at the descriptions on pages 6–7 to find out. For example, the Freshwater Shrimp (left) is a crustacean. Each animal is identified by a special sizing symbol (see the front of the book to view them).

3 **Look through the pages of animals with your habitat picture band at the top.** The picture and information given for each animal will help you to identify it.

4 **If you can't find the animal there,** look through the other sections. Animals move around, and you may see them in more than one habitat. You will find the frog (right) is a Spotted Frog (see page 49).

5 **If you still can't find the animal,** you may need to look in a larger field guide (see page 79 for some suggestions). You may have spotted a very rare creature!

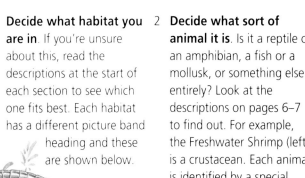

Finally, the nymph crawls up a plant stem into the air, and moults for a final time. Out crawls the adult dragonfly, with its brightly colored body. It waits for its wings to dry, then flies off, a land-based insect.

Watch for empty skins attached to waterplants

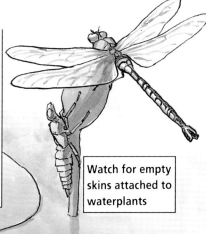

Top-of-page picture bands

This book is divided into different freshwater habitats. Each habitat (type of countryside) has a different picture band at the top of the page. These are shown below.

Found Almost Everywhere

Ponds & Shallow Lakes

Swamps & Marshes

Deep & Cold Waters

Fast Streams & Rivers

Slow Rivers & Canals

What To Look For

Reptiles

Alligators, turtles, snakes, and lizards are all reptiles. Lizards live on land and are not covered in this book. Reptiles have skin covered in scales, or hard plates and shields. They have claws on their toes.

Amphibians

Most frogs have smooth skin that is wet to the touch

Toads have tough, warty skin that is dry to the touch

Salamanders have smooth skin; most have stout front and back legs of equal size. The belly (underneath) is a different color to the back.

Frogs, toads, newts, and salamanders are all amphibians. They have moist skin with no scales. They do not have claws on their toes. They lay jelly-like eggs in water which hatch into tadpoles.

Fish

It is usually easy to recognize a fish. However, be careful not to mistake eels for amphiumas or sirens—both are amphibians, and have tiny legs. Look for the number and position of the fins, especially those on the back which are called the dorsal fins.

Eels have one long fin

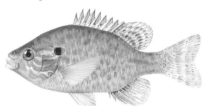

Sunfish have one long fin, spiny in front and soft at the back

Minnows have one short fin

Perch and darters have two fins, close together

Trout have two fins, one large, one tiny

Mollusks

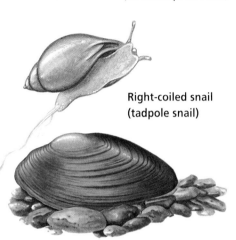

**Left-coiled snail
(common pond snail)**

**Right-coiled snail
(tadpole snail)**

Two shells (pearl mussel)

All mollusks are soft-bodied animals, and most have a shell. Animals with one shell are called gastropods. Those with two hinged shells are called bivalves. When you see a water snail, check whether it coils to the left or right. Hold the shell upright, with the aperture facing you and follow the coil with your finger.

Crustaceans

Crustaceans have a hard, plated skin, long feelers for finding food, and many pairs of jointed legs. As they grow, they must molt their skin and grow a bigger one. Crayfish, shrimps, and prawns are all crustaceans.

Insect larvae

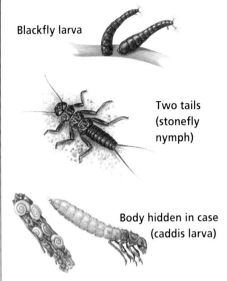

Blackfly larva

**Two tails
(stonefly
nymph)**

**Body hidden in case
(caddis larva)**

Most of the insect larvae in this book are nymphs. They have bodies that are divided into three parts: head, middle (thorax), and rear (abdomen). There are three pairs of jointed legs on the thorax. There are no legs on the abdomen though some have many pairs of feathery gills— try to count these.

A fly larva looks very different to a nymph. It has a small body divided into segments often with tufts of hair, and a small head.

Worms have no legs and no hard skin or shell. They are usually long and thin. Flatworms have flat, soft bodies and glide over rocks. A true worm's body is divided into segments, and there is no obvious head. A leech's body is divided into segments. There are suckers at both ends, but they are difficult to see.

Worms & worm-like animals

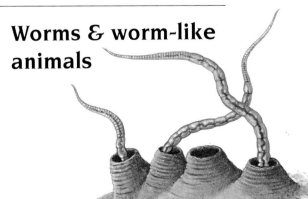

Found Almost Everywhere

Some water animals don't seem to mind where they live. They can be found in streams, rivers, ponds, lakes, or even in swamps. Many of these creatures are also very wide-ranging. You may spot them throughout much of North America and even in other parts of the northern hemisphere if you should take a vacation in Europe. This picture shows ten animals from this section; see how many you can identify.

American Eel, Leopard Frog, Bluntnose Minnow, Three-spined Stickleback, White Sucker, American Toad, Painted Turtle, Snapping Turtle, Brown Water Snake, Sludge Worms.

Reptiles

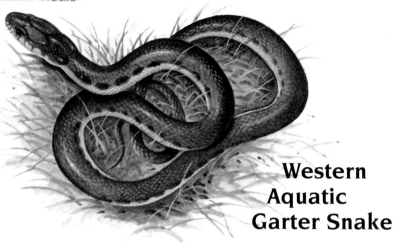

Western Aquatic Garter Snake

You may find this snake hard to identify because it varies in color. It can have three stripes, or be spotted or blotched. Sometimes it has no back stripe at all. It lives mainly in rivers and streams, but can also be found in coastal marshes and in high mountain waters. This snake is active mainly during the day, when it feeds on fish, frogs, toads, salamanders, earthworms, and leeches. Instead of laying eggs, females of this type of snake give birth to live young.

Group: Colubrid Snakes (Colubridae) – Size: 18–60 ins
Distribution: Western U.S, from Oregon to northern Baja California, California, and east into western Nevada

Brown Water Snake

This snake is brown or dark brown and has quite a stout body. There are three rows of big, dark blotches running down its back and sides, and its yellow belly has dark spots. You will find the Brown Water Snake around rivers, lakes, and large ponds, often taking the sun on banks or on tree branches that overhang the water. It is most active by day, and feeds on frogs and fish. Beware of this snake, as it is aggressive and gives a nasty bite.

Group:
Colubrid Snakes (Colubridae)
Size: 28–70 ins
Distribution: Southeastern U.S. from Virginia coast through Florida to southwestern Alabama

Northern Water Snake

This snake is brownish in color with dark bands in the neck area and squarish dark blotches on its back and sides. Underneath, its belly is white, yellow, or gray. This snake can be seen in most watery places, such as lakes, ponds, rivers, ditches, and bogs. It hunts by day and by night, catching and eating frogs, salamanders, turtles, and small fish. During the day, it basks on rocks. The Northern Water Snake is more likely to flee than attack, but if angry, it will strike with its body and bite. Unfortunately, water snakes are often mistaken for poisonous cottonmouths and are killed.

Group: Colubrid Snakes (Colubridae) – Size: 22–53 ins
Distribution: Southeastern Canada and most of eastern U.S. excluding Florida panhandle

Diamondback Water Snake

So called because of the diamond pattern on its back, this snake is greenish brown with black markings and a yellow belly. The Diamondback Water Snake is active during the day, and can often be spotted basking on logs along the edges of lakes, rivers, streams, or in swamps or ditches. It feeds on fish and frogs. Be careful if you spot this snake, as it is aggressive and quick to bite.

Group: Colubrid Snakes (Colubridae)
Size: 30–64 ins
Distribution: South central U.S. from Illinois and Indiana, south to Mississippi and Alabama, west to Texas and Mexico, north through Oklahoma, Kansas, and north Missouri

Painted Turtle

This little turtle gets its name from its brightly patterned shell, which looks as if it has been painted with red and yellow bars. The basic color of the shell is olive or black, and it is smooth and rather flattened. The Painted Turtle's neck, legs, and tail match its shell beautifully, as they are also patterned with red and yellow stripes. Painted Turtles live in shallow streams, rivers, and lakes. They sunbathe on logs, often in big groups. Their favorite foods include small fish, tadpoles, frogs, and water plants. They are the most widespread turtles in North America.

Group: Pond and Box Turtles (Emydidae) – Size: 4–10 ins
Distribution: Southern Canada and most of north and east U.S., British Columbia to Nova Scotia, south to Georgia, west to Louisiana, north to Oklahoma, and northwest to Oregon

Snapping Turtle

This aggressive turtle is one to keep away from.
You will recognize it easily by its big, knobbly head, hooked jaws, and its tail which is as long as its shell. The Snapping Turtle will snap at just about anything, including your fingers and toes! It eats plants and a great variety of live and dead animals. The shell is plain tan to dark brown and serrated towards the end. The undershell is yellow to tan and quite small. You'll rarely see the Snapping Turtle out of water, or sunbathing on the riverbank. It prefers to rest in warm, muddy, shallow water, with only its snout visible, waiting for its next meal.

Group:
Snapping Turtles (Chelydridae)
Size: 8–18½ ins
Distribution: South Alberta to Nova Scotia, and south to the Gulf

Amphibians

American Toad

Toads can be difficult to tell from each other, but unlike most frogs, they have a rough, warty skin. The large American Toad can be brown, brick-red, or olive, with patterns in lighter colors. Warts on its tough hide are brown to orange-red, and there are large bony crests above its eyes. There may be a light stripe down the middle of its back, and its belly usually has spots. The American Toad lives in a variety of places, from grassy backyards to mountains, wherever there is enough moisture and enough insects for it to eat. It usually becomes most active at night. In the spring, listen for the American Toad's beautiful voice, which sounds like a musical trill.

Group: Toads (Bufonidae)
Size: 2–4½ ins
Distribution: Widespread in eastern North America: Maritime Provinces to southeast Manitoba, south to Mississippi and northeast Kansas

Western Toad

This toad is large and green in color with a light stripe down the middle of its back. Warts on its body are reddish and surrounded by black blotches. It has no crests above its eyes, which may help you to identify it. The Western Toad is found in various places, from desert streams and springs to woodlands, grasslands, and mountain meadows. You will see it walking rather than hopping. It makes burrows for itself or inhabits those of squirrels and other small mammals. Listen for the voice of the Western Toad at twilight. It sounds rather like a chick chirping.

Group: Toads (Bufonidae) – Size: 2½–5 ins
Distribution: Pacific coast from south Alaska to Baja California, east to Alberta, Montana, Wyoming, Utah, Colorado, and Nevada

Mud Puppy

Group: Mud Puppies and Waterdogs (Proteidae)
Size: 8–17 ins
Distribution: Most of eastern half of the U.S. excluding coastal strip. Also southeast Manitoba to south Quebec

The Mud Puppy or Waterdog is a kind of salamander that remains a larva all its life. You can easily recognize the Mud Puppy by the maroon gills around its neck, that look like a feathery collar. It is gray to rusty brown in color, with faint, blue-black spots. The belly is gray with dark spots. It has four toes on each of its four feet. It lives in all kinds of lakes, rivers, and streams, including those that are muddy and full of weeds. It feeds on fish, crayfish, water insects, and mollusks.

Chorus Frog

You are more likely to hear this treefrog than see it. During the spring, it sings night and day in grassy areas near water, in woodlands, and in swamps, but it is always well hidden in the grass. The sound it makes is like the sound you get by running a fingernail over the teeth of a comb. The Chorus Frog has greenish brown, smooth skin, with three dark, sometimes brown, stripes down its back. There is a dark stripe through its eye, and a white stripe along its upper lip.

Group: Treefrogs (Hylidae)
Size: 3/4–1 1/2 ins
Distribution: Widespread over most of North America from Alberta to New York, south to Georgia, west to Arizona

Northern Leopard Frog

This frog is easy to identify, with its brown or green skin covered with large spots. The spots are rounded and have light-colored borders. The Northern Leopard Frog lives in marshes, moist fields, and mountain meadows. If you spot one, it will probably leap in a zigzag pattern to the safety of the nearest water. Its voice is low, and sounds like a snore, followed by clucking noises. Listen for this noise at night in particular. It is found throughout northern U.S., except on the West Coast. The Southern Leopard Frog has a longer, pointed head, and only a few spots on its sides. It is active mainly at night, and its voice consists of throaty croaks.

Group: True Frogs (Ranidae)
Size: 2–5 ins
Distribution: (Southern) From New York to Florida Keys, west to Texas and eastern Oklahoma, north to eastern Kansas

Woodhouse's Toad

This toad has a light stripe on its back, obvious bony crests above its eyes, and dark spots with warts. Its general color is yellow to green to brown. It lives in many areas, including marshes, river bottoms, desert streams, backyards, and rain puddles. Woodhouse's Toad becomes most active at night, and may be spotted catching insects that are attracted to lights. It makes a sound rather like a lamb bleating in the distance.

Group: Toads (Bufonidae)
Size: 2 1/2–5 ins
Distribution: Common throughout most of U.S.

Fish & Worms

Three-spined Stickleback

This little fish gets its name from the three sharp spines on its back. There are no scales on its body, just a row of bony plates along the sides. The top of its body is brownish olive, and underneath it is white or silvery: males have red bellies during the breeding season. The Three-spined Stickleback is not fussy about where it swims, and lives in both salt and fresh water. It particularly likes shallow water, where there are weeds growing. Look out for Stickleback nests in the spring. The males make them out of water plants and chase away other fish from their territory with their sharp spines. Three-spined Sticklebacks feed on young shrimps, small fish, and fish eggs, as well as on plants.

Group: Sticklebacks (Gasterosteidae)
Size: Up to 2 ins (exceptionally 4 ins)
Distribution: Pacific Coast from Baja California north to Alaska, Atlantic Coast from Chesapeake Bay north to western shore of Hudson Bay and Baffin Island

Bluntnose Minnow

You can see how this little fish got its name. Its snout is blunt instead of pointed. The scales on its body are a silvery color outlined in black to make a crisscross pattern. There is a strong black line running along its side. It has one short fin on its back. This minnow lives in several habitats, from small brooks to large lakes. It prefers water that is full of plant life, and feeds on a variety of plants, as well as water insects and their larvae. Females lay their eggs between spring and fall under logs or stones.

Group: Carps and Minnows (Cyprinidae)
Size: Up to 3 ins
Distribution: Eastern North America from Quebec west to Manitoba, New York south to Georgia, through Great Lakes and Mississippi Valley

White Sucker

This fish gets its name from its sucker-like mouth, which has very obvious fleshy lips. Its body is olive-brown, with a silvery-white belly. It has a single short back fin. The White Sucker lives in many different types of water, from fast-flowing to slow-moving, from weedy to clear. It creeps along the slimy bottom mud, feeding on insect larvae, mollusks, crustaceans, and algae. During the spring breeding season, the male changes color. Its back becomes lavender-colored, and a red band appears along its side.

Group: Suckers (Catostomidae)
Size: Up to 18 ins
Distribution: North America east of Rockies from Hudson Bay in the north to Oklahoma, northern Mississippi and northern Georgia in the south

Spotted Flatworm

Like all flatworms, this creature has a thin, flat, soft body. It has a triangular-shaped head and two simple, black eyes. As the name suggests, it has spots and streaks all over its body. Its mouth is on the underside, and it crawls along on muddy bottoms of ponds, lakes, slow rivers, and streams, and feeds on decaying matter there. Flatworms look like blobs of jelly out of water. They will expand and glide around if put in a dish of water.

**Group: Flatworms (Turbellaria) – Size: Up to ¾ in
Distribution: Common across North America**

American Eel

You'll recognize this, and all eels, by their snakelike shape and fins which run continuously along the long body. The American Eel is sometimes called the Silver Eel because young ones have silver bellies. It has a pointed snout and a wide mouth. It prefers muddy water, but can be found in almost any type of water. Like other freshwater eels, it undertakes a long journey after living in fresh water for a few years. It travels down to the sea, to an area south of Bermuda, where it spawns then dies. The eggs hatch into larvae, which drift back to the North American coast. The larvae change into young eels or "elvers," which enter the freshwater streams and rivers. Then the cycle starts all over again.

**Group: Eels (Anguillidae) – Size: Up to 30 ins
Distribution: Atlantic and Gulf Coasts and
coastal rivers of North America**

Flatworm

There are many different kinds of flatworm and most are difficult to identify. This flatworm is blackish in color and has a pointed head. It lives in springs and headwaters. It creeps along the bottom and feeds on dead and decaying matter. You can collect flatworms for further study by putting some pieces of raw meat into the water. After a while, the meat can be removed and the worms shaken off into a container of water.

**Group: Flatworms (Turbellaria)
Size: Up to ¾ in
Distribution: Common across the U.S.**

Sludge Worms

Sludge or Mud Worms are long, thin, and reddish, a color which comes from the blood. These worms live in the mud at the bottom of lakes, stagnant ponds, or polluted rivers. They are able to survive in poor conditions because they do not need much oxygen. They build soft mud tubes in which they live head down, with their tails waving about in the water. You may see them in aquarium shops, where they are sold as fish food.

**Group: Tubifex Worms (Tubificidae) – Size: Up to 2 ins
Distribution: Widespread
throughout North America**

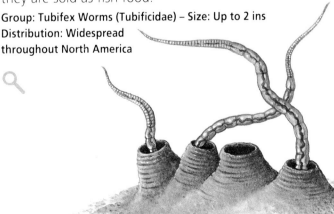

Mollusks

Spire Snail

These tiny snails have a shell with four to five whorls. They may be reddish brown, grayish brown, or tan in color. The snails live in many kinds of freshwater habitats among thick weeds. They prefer unpolluted water. They can survive out of water by closing the shell opening with a horny plate called an "operculum." They often occur in huge numbers.

Group: Spire Shells (Hydrobiidae)
Size: Up to 1/4 ins
Distribution: Widespread throughout North America

Flat-ended Spire Snail

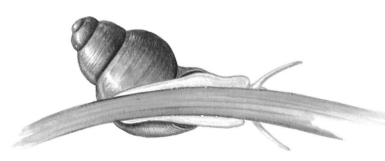

This type of Spire Snail is quite wide and solid. It has about five turns, or "whorls," but the first two are sunken below the third, which is the biggest. The "spire" at the top is barrel-shaped. This snail lives in lakes, ponds, and rivers of any size, and tends to stay near plant life or on sandy or muddy bottoms. Its body is white, and it has an operculum which it uses to seal the opening of its shell.

Group: Spire Shells (Hydrobiidae) – Size: Up to 1/4 ins
Distribution: Quebec to western Northwest Territories and Alberta, and from New York to Iowa, Arkansas, and Kentucky

Pea Clams

Also called Pea-shell Cockles, these little creatures have shells in two parts, so they are bivalves. The shells are yellowish or buff in color, and rounded in shape. They can be found in many watery habitats, even water troughs, and sometimes burrow into gravel or sand at the bottom. They are difficult to tell from Fingernail Clams. Left in a bowl of water, they may send out two short, joined siphon tubes, through which they draw in water.

Group: Orb Mussels (Sphaeriidae)
Size: Up to 1/4 ins
Distribution: Found throughout North America

Fingernail Clams

Also called Orb-shell Cockles, these bivalves are yellowish or greyish brown and rounded, but are slightly bigger than Pea Clams. They burrow into the bottom of various freshwater habitats. Young ones may clamber about on water plants. If you collect some and put them into water, you may see the two siphon tubes and the strong foot.

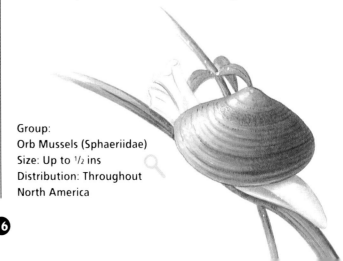

Group:
Orb Mussels (Sphaeriidae)
Size: Up to 1/2 ins
Distribution: Throughout
North America

Eastern Lamp Mussel

This mussel has a wide, dark brown to black shell covered with wide, raylike marks. It gets the name "Lamp Mussel" because the rays look like light rays. The Lamp Mussel is common in rivers and lakes of all sizes, where it lives on gravel, sandy, or muddy bottoms, filtering plankton from the water.

Group: Pearly Mussels (Unionidae)
Size: Up to 4 ins
Distribution: Lower St. Lawrence system, south to Pee Dee River system in South Carolina

Fat Mucket

The Fat Mucket is similar to the Lamp Mussel, but its shell is yellowish, greenish, or brownish, and the rays are narrow, instead of wide. This bivalve lives in rivers and lakes of all sizes. The young attach themselves to various types of fish, including White Bass, White Crappie, and Yellow Perch. Here they feed for a few weeks, before dropping to the bottom of the water.

Group: Pearly Mussels (Unionidae)
Size: Up to 1/2 ins long
Distribution: Canada, upper Ohio to the Mississippi, and from New York south to Minnesota and Arkansas

Filter Mussel

This mussel has a shell in two parts, so it is a bivalve. Each shell is oval in shape, and the outside is rough in texture and brown or black in color. Inside, the shell has a pearly coating. The Filter Mussel lives on the bottom of shallow ponds, rivers, lakes, and streams. Like all mussels, the Filter Mussel sucks in water through a tube, or siphon. It then filters out food, usually plankton, and ejects the waste water through another siphon. Young Filter Mussels attach themselves to the skin of fish and feed off them for the first few weeks of their lives. This does not harm the fish.

Group: Pearly Mussels (Unionidae) – Size: 3–4 ins
Distribution: Southeastern Canada and East Coast U.S. to Georgia. Similar species in other parts of North America

A Collecting Trip

You don't need much equipment, and you certainly don't need to be an expert naturalist to find and study water creatures. Wear old clothes and rubber boots or old sneakers, as the area around the pond may be quite muddy. Don't forget to take this field guide! When you get there, don't run or stamp around on the bank, and don't shout. The noise will frighten animals away.

What to take

You may also find the following things useful:

1 **Dipping net** (buy one from a pet store).
2 **Underwater viewing box** (see right for how to make it).
3 **Empty, white margarine or ice cream container** for watching your catch.
4 **A pipette (eye-dropper) or paintbrush** for moving small animals without hurting them.
5 **Glass or plastic jars with screw-on lids** for carrying species home.
6 **Your field note book, with pens and pencils** for note-taking.
7 **A magnifying glass**. Buy one labeled x4 or x6 and wear it on a string around your neck.
8 **A camera** to record the various sites.
9 **A lightweight backpack** to carry everything in.

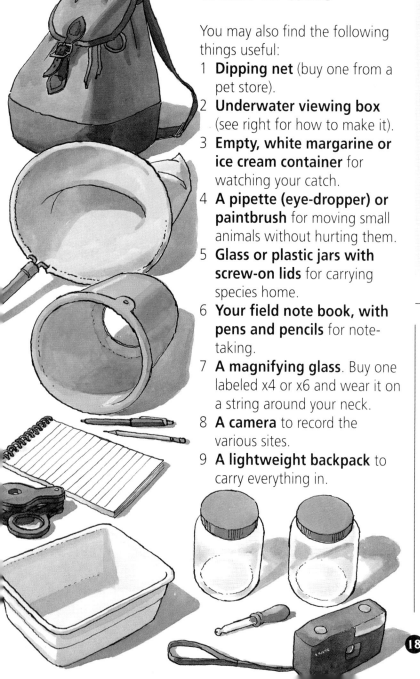

SAFETY DOs and DON'Ts

1 **Always tell an adult** when you are going on an expedition to a pond or river.
2 **Always take a friend**.
3 **Always test the depth of water with a long pole** before going wading—it may be deeper than you think.
4 **Always test any log or stone** before you use it as a stepping stone.
5 **Always wear a life jacket** if you are going out in a boat—even if you can swim well.
6 **Don't lean too far over the water's edge**.
7 **Don't climb on tree branches overhanging** the water—they might break.

Underwater viewing box

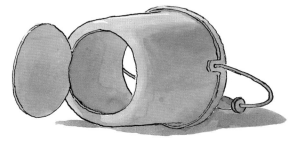

Making your own viewing box for the water will allow you to watch water creatures and their activities in their natural habitat.

1 **Find a plastic container** such as an old bucket or wastepaper basket. Get an adult to cut the bottom out leaving a rim about an inch wide.

Handling amphibians

You may be able to catch newts, young salamanders, frogs or toads using a dip net in a pond (see page 33). Most salamanders, newts and frogs can be safely handled. But remember, amphibians are "cold-blooded" and have soft, moist skins. They do not like being held in warm, dry hands. Tip them into a plastic tub with some damp vegetation. When you have finished looking at them, put them back where you found them.

Some toads and a few salamanders can secrete an unpleasant liquid from their skins. This can cause swelling and discomfort if it enters a cut or gets in your mouth or eyes. **Rinse the affected part with clean water as soon as possible if this happens.**

Keeping records

Making notes and quick drawings in your field notebook will help you identify fast-moving animals. Transfer your notes into a ring binder when you get home and record each trip on a separate sheet of paper. Stick in your sketches, photos, and any other samples you may have collected during your expedition.

1 **When you visit a new site,** give it a name, write down the date and what sort of habitat it is (pond, river, marsh, etc.).
2 **Each time you visit that site,** record what the weather was like and what time of day it was.
3 **When you see an animal,** watch it carefully. Make a note of its shape, color and size to help identification.
4 **Write down what the animal was doing** (swimming, eating, resting, etc.), how many there were, and if it was making a noise.

2 **Get an adult to cut a piece of clear plastic** to fit inside the container.
3 **Stick it down onto the rim** with waterproof adhesive—you can buy this from a pet store.
4 **Take your viewing box along to a pond,** lake, or reservoir. Carefully lean over the bank and place the clear plastic end of your box in the water. Look through the hole for snails crawling, fish swimming, and other creatures.
5 **You can also wade upstream** in shallow streams and rivers, wearing your rubbers. These areas are good for finding mussels. A sunny day is best.

Ponds & Shallow Lakes

All ponds have still, shallow water. Shallow water means that light can penetrate to the bottom, allowing water weeds to grow. The weeds provide food and shelter for many animals. Because the water is still, there is not very much oxygen in it, and it may be stagnant near the bottom. Only animals that can put up with this can live in ponds.

Many pond snails, and all frogs, toads, newts, and reptiles can live in stagnant water because they come to the surface to breathe. Others, such as mud puppies and tadpoles, have very large gills to help them take enough oxygen from the water.

Most ponds are made by people. Small ponds often have to be looked after to stop sedges, rushes, and other swamp plants from growing right into them and filling them up. Many animals found in ponds and lakes will also be found in slow-moving streams and rivers (see pages 72–77) where conditions are very similar to a pond.

As many natural wetlands are now being destroyed, ponds are becoming more and more important as refuges for water animals. This picture shows sixteen animals from this section; see how many you can identify.

Alderfly nymph, Diving Beetle adult & larva, Whirligig Beetle, Bullfrog, Caddisfly nymphs, Damselfly adult & nymph, Dragonfly adult & nymph, Fish Leech, Snail Leech, Water Louse, Fathead Minnow, Red-spotted Newt, Red Shiner, Dwarf Siren, Great Pond Snail, Longear Sunfish.

Reptiles

Chicken Turtle

You will recognize this turtle by its long, stripy neck that is nearly as long as its shell. The shell itself is green or brown, with a yellowish netlike pattern, and an orange or yellow border. The Chicken Turtle lives in shallow ponds and lakes that are filled with plants, or in ditches and swamps. It feeds on small water creatures such as tadpoles. You may see it basking in the sun, or crossing highways. Unfortunately, it is sometimes run over by cars.

Group: Pond and Box Turtles (Emydidae) – Size: 4–10 ins
Distribution: Southeastern U.S. coastal plain from Virginia to Florida, west to eastern Texas, north to Oklahoma and Missouri

Spotted Turtle

This small turtle is easy to recognize with its smooth, black shell covered with round, yellow spots. There are also spots on its head, neck, and legs. The undershell is creamy yellow with black blotches around the edge. Females have orange eyes, while males' eyes are brown. Look for them in marshy meadows and boglands, as well as in ponds and shallow, muddy-bottomed streams. It likes to sunbathe in the spring, on a log at the water's edge. Favorite foods include insects and water plants.

Group: Pond and Box Turtles (Emydidae) – Size: 3½–5 ins
Distribution: Coastal plain of eastern U.S. from Maine to north Florida. Also in states bordering the south of the Great Lakes.

Western Pond Turtle

This turtle has a smooth, flattish, olive-brown shell with darker flecks. It lives in weedy ponds, marshes, rivers, streams, and ditches, where it feeds on plants, insects, worms, and fish. It likes to sunbathe out of the water, but will dive back in if you disturb it.

Group: Pond and Box Turtles (Emydidae) – Size: 3–7 ins
Distribution: West coastal strip from British Columbia to Baja California

Pond Slider

This turtle's other name, the Red-eared Slider, comes from the red blotches behind each ear. These can be yellow, as shown here. The turtle's shell is oval in shape, olive-brown in color, and patterned with yellow marks. It lives in slow rivers, shallow streams, swamps, and ponds with lots of plant life and soft bottoms. It can swim quite fast, kicking its powerful back legs. Look for groups sunbathing on logs near the water's edge. Young Pond Sliders feed on water insects, mollusks, and crustaceans. When older, their diet becomes vegetarian.

Group: Pond and Box Turtles (Emydidae) Size: 5–12 ins Distribution: Southeastern U.S. from Virginia to northern Florida, west to New Mexico, south to Brazil

Florida Softshell

The Florida Softshell has no hard shell, and looks rather like a brownish grey, leathery pancake! It also has an unusual tubelike snout and a long neck. The soft shell is thickened all around the edge and has small bumps on the front part. Look for this turtle swimming in sandy-bottomed lakes, ponds, and canals, with just its snout visible. It also likes to bask in the sun on banks or logs. It feeds on crayfish, snails, frogs, and fish. Be careful if you handle a Softshell as it has surprisingly sharp jaws.

Group: Softshell Turtles (Trionychidae) – Size: 6–12 ins
Distribution: South Carolina southward

Mud Snake

This big, black, shiny snake has triangular, pink, or red bars on either side and a red belly. Smooth scales cover the body, and there is a sharp spine at the end of its tail, which is not poisonous. The Mud Snake lives around swampy lakes and slow-flowing, muddy-bottomed streams. On rainy nights you may see it crossing roads in swampy areas. It feeds mostly on amphiumas and "sirens" (see pages 26–27).

Group: Colubrid Snakes (Colubridae) – Size: 36–80 ins
Distribution: Southeastern U.S. from Virginia to southern Florida, west to eastern Texas, and north in Mississippi Valley to southeastern Illinois

Ribbon Snake

A slim, stripy snake, the Ribbon Snake is olive brown where it is not striped. The stripes are thin and usually orange-tan and cream. It lives around weedy lakes and also in marshes, ditches, streams, and rivers. If frightened, it slithers into the water and glides away across the surface. It feeds at the water's edge on frogs, tadpoles, and small fish. Female Ribbon Snakes give birth to live young.

Group: Colubrid Snakes (Colubridae)
Size: 20–48 ins
Distribution: Eastern half of U.S.

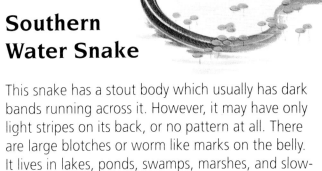

Southern Water Snake

This snake has a stout body which usually has dark bands running across it. However, it may have only light stripes on its back, or no pattern at all. There are large blotches or worm like marks on the belly. It lives in lakes, ponds, swamps, marshes, and slow-moving streams. It becomes active at night, when it feeds on frogs, tadpoles, and fish. Females give birth to live young.

Group: Colubrid Snakes (Colubridae) – Size: 16–60 ins
Distribution: Southern U.S. from North Carolina to Florida Keys, west to eastern Texas, north in Mississippi River Valley to southern Illinois

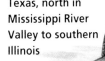

Bullfrog

This frog is hard to miss. It is the biggest frog in North America and is colored green or yellow with dark gray mottling on its back. Its belly is cream, and is often also mottled with gray. The round eardrum behind the eye is very large. The Bullfrog lives in lakes, ponds, bogs, and slow-flowing streams. It can often be seen sitting at the water's edge, but will quickly flee if anyone comes along. Favorite foods include insects, crayfish, other frogs, and minnows. It is most active at night. Listen out for the Bullfrog's loud voice. It sounds as if it is saying "JUG O' RUM."

Group: True Frogs (Ranidae)
Size: 3–8 ins
Distribution: Eastern and central U.S., New Brunswick and parts of Nova Scotia, introduced in the West

Red-legged Frog

This frog gets its name from the red color on the underside of its hind legs. Its back is reddish brown to gray, with darker specks and blotches, and its belly is yellow, fading into red. It usually also has a dark patch over and behind the eye with a white jaw stripe. The Red-legged Frog lives near ponds and lakes that are full of plants. It also likes damp forests and woodlands. It is most active during the day, and its voice consists of a series of throaty notes.

Group: True Frogs (Ranidae)
Size: 2–6 ins
Distribution: Western coastal strip from Vancouver Island, to northern Baja California

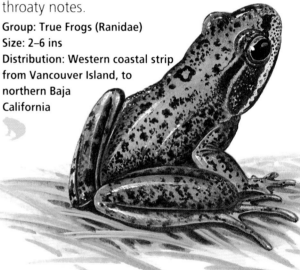

Spring Peeper

The little Spring Peeper can be tan, brown, or gray, with a dark cross on its back. It lives in woodland areas, near ponds or swamps, and becomes most active at night. The name comes from the high-pitched whistle it makes, which sounds like the jingling of bells when a group sings together. This sound is one of the first signals that spring has arrived. Many other tree frogs can be found in trees near water.

Group: Tree Frogs (Hylidae) – Size: 3/4–1 1/2 ins
Distribution: Most of southeast Canada and eastern half of U.S.

Northern Cricket Frog

Group: Tree Frogs (Hylidae)
Size: 3/4–1 1/2 ins
Distribution: Most of eastern U.S. except Florida panhandle

This knobbly-skinned frog is colored greenish brown, yellow, red, or black. Look for a dark triangle between its eyes and a dark stripe on its thigh. Its legs are quite short, but this doesn't stop it hopping. This frog lives in ponds with shallow water and lots of plants, and also by slow-flowing streams. It is active during the day, and there are often many of them around, but you won't catch one because they are too fast at hopping away. The Northern Cricket Frog's voice sounds like a series of clicks, similar to a cricket.

Green Frog

Although it is called a Green Frog, this frog can be brown or bronze in color as well. Two raised ridges run down both sides of the back. Its belly is white, with a pattern of lines or spots. Males can be identified by their yellow throat. The Green Frog can be seen in shallow fresh water, from springs, swamps, and brooks, to edges of ponds and lakes. It usually becomes active at night, and makes a noise like the twang of a loose banjo string.

Group: True Frogs (Ranidae)
Size: 2–4 ins
Distribution: Common throughout eastern North America

Many-lined Salamander

This salamander is so called because of the many dark lines along its sides. Its overall color is brown or dull yellow, and it has a yellow belly with specks. It also has a small head and a short tail. Look for this salamander in ponds, slow streams, and swampy areas. It is usually in the water, but sometimes sits under logs where the ground is damp.

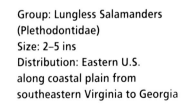

Group: Lungless Salamanders (Plethodontidae)
Size: 2–5 ins
Distribution: Eastern U.S. along coastal plain from southeastern Virginia to Georgia

Amphibians

Red-spotted Eastern Newt

This newt is yellowish brown or olive-green with black-bordered red spots on its back. At a younger stage, it is called an "eft" and is bright orange-red in color. The adult newt lives in ponds, small lakes, marshes, and streams, and in damp woodlands nearby. It feeds on worms, insects, crustaceans, and mollusks. The efts are very bold, and may be seen in huge groups on the forest floor after a summer shower.

Group: Newts (Salamandridae) – Size: 2$\frac{1}{2}$–5$\frac{1}{2}$ ins
Distribution: Nova Scotia west to the Great Lakes, south to northwestern South Carolina, central Georgia, and Alabama

Broken-striped Eastern Newt

The Broken-striped Newt is so called because of the broken black-bordered stripe that runs down its back. It is yellowish brown or olive-green in color. The young, land-dwelling eft, shown here, is reddish brown, and has red stripes with a less strong border than the adult's. The adult newt lives in pools, ponds, ditches, and quiet parts of streams. The efts can be found under logs in damp places.

Group: Newts (Salamandridae) – Size: 2$\frac{3}{4}$–5$\frac{1}{2}$ ins
Distribution: Coastal plain, northeastern North Carolina and southeastern South Carolina. Other varieties of the Eastern Newt can be found throughout the eastern U.S.

California Newt

This newt has rough skin which is tan to reddish brown on the back, and yellow to orange on the belly. The eyes are large, with light-colored lower lids. When breeding, males become smooth-skinned. The California Newt lives in quiet streams, ponds, and lakes, and in surrounding evergreen and oak forests. In a rainy season, you may see this newt by day, but when it is dry, it burrows under moist forest leaf litter. When threatened, it reveals its brightly colored belly to frighten off predators.

Group: Newts (Salamandridae)
Size: 5–8 ins
Distribution: Coastal California

Rough-skinned Newt

It is difficult to tell this newt from the California newt. It has even rougher, warty skin, which is light brown to black on its back, and bright yellow or orange on its belly. Its eyes are small, with dark lower lids. When breeding, males develop smooth skin. The Rough-skinned Newt lives in ponds, lakes, and streams that have plant life, and also in surrounding moist woodlands. It is very fond of the water, but you may see it wandering on land on humid days. If threatened, it assumes a warning pose to frighten off its attacker.

Group: Newts (Salamandridae)
Size: 5–8$\frac{1}{2}$ ins
Distribution: Pacific Coast, from Santa Cruz county to southeastern Alaska

Dwarf Siren

Like all sirens, this slender, eel-like creature never gets past the larva stage of its life. It is brown or light gray, with light stripes on its sides. You can see the gills close to the head. Called the Dwarf Siren because it is the smallest siren of all, this creature lives in ditches, swamps, and weedy ponds. You may find it hard to spot one, as they hide among the water plants.

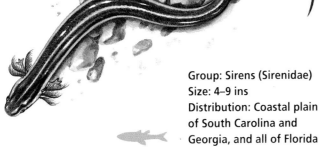

Group: Sirens (Sirenidae)
Size: 4–9 ins
Distribution: Coastal plain of South Carolina and Georgia, and all of Florida

Siren

Sirens are salamanders that live in quiet, weedy waters and remain larvae all their lives. With their long, slender bodies, they look rather like eels. They have visible gills at the neck, no hind legs, and tiny front legs. They feed at night, mainly on small animals. If the water in their home dries up, they burrow into the mud and survive by covering themselves with a sticky cocoon which keeps in the moisture.

Group: Sirens (Sirenidae)
Size: 7–36 ins
Distribution: Coastal plain of southeastern U.S. from Virginia to Texas, and all of Florida

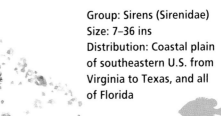

Amphiuma

These are another type of salamander. They have long, slender bodies and look rather like eels. They have four tiny legs that are of no use, and each leg has one to three toes depending on the species. Amphiumas do not change much from the larva form. They lose their gills, but the gill slits can still be seen. They are active at night, and feed on creatures such as crayfish, frogs, snakes, fish, and even other Amphiumas. They can be seen in muddy ditches, ponds, swamps, and streams. Be careful if you find one because they give a nasty bite.

Group: Amphiumas (Amphiumidae)
Size: Up to 45 ins
Distribution: Coastal plain of southeastern U.S. from Virginia to Texas, and all of Florida

Fish

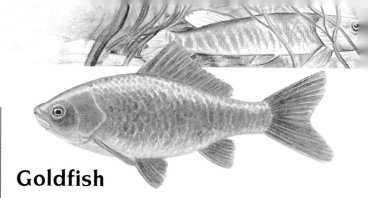

Red Shiner

This fish is a brilliant steel-blue color with large, diamond-shaped scales. It has an arched back, and red fins, except for the back fin. The Red Shiner lives in ponds, slow-flowing streams, and large rivers which flow over sand or gravel. It feeds on bits of water plants, small insects, and crustaceans. In the early summer, females lay their eggs on plants under the water. Another name for this fish is the African Fire Barb. It is often called this name when sold as an aquarium fish.

Group: Carps and Minnows (Cyprinidae)
Size: Up to 3 ins
Distribution: Central U.S., especially Plains region west of the Mississippi River from Wyoming to Minnesota, south to Texas

Channel Catfish

This catfish can be recognized by its spotty body and forked tail. It is a slaty-brown color above and silvery-white below. Long barbels surround its mouth and, like all catfish, its skin is smooth with no scales. The Channel Catfish lives in ponds, lakes, and reservoirs or large rivers that flow over gravel or sand. It feeds on other fish, insects, mollusks, and crayfish, mainly at night. Females lay their eggs under a stone or in a nest below a stream bank and the male guards the nest.

Group: Catfishes (Ictaluridae) – Size: Up to 48 ins
Distribution: Found across eastern North America from the Great Lakes south to the Gulf of Mexico, and into Mexico

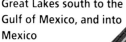

Goldfish

You've probably seen this fish more in fish bowls than in the wild, but it has been introduced into warmer waters of the U.S. As its name suggests, the Goldfish is a goldish orange color, with scales that are easy to see. It has a long back fin which has one serrated spine. This fish likes warm water that has plenty of plants in it, where it feeds on water insects, mollusks, and crustaceans, as well as plants.

Group: Carps and Minnows (Cyprinidae)
Size: Up to 16 ins
Distribution: Warmer waters of the U.S., Canada, and Mexico

Common Carp

This Carp can be recognized by its fat body, its scales which give a crisscross pattern, and its long back fin. At the top of this fin is a hard, serrated ray, but the rest of the rays are soft. The Carp is a golden-olive color and has two short barbels on each side of its jaw. It prefers to live in warm water and can be seen in quiet ponds, lakes, and sluggish rivers which have plants in them. Favorite foods include mollusks, crustaceans, and insect larvae, as well as algae and plants. Females lay eggs in the spring, which stick to plants, or sink to the bottom.

Group: Carps and Minnows (Cyprinidae)
Size: Up to 30 ins
Distribution: Southern Canada and throughout the U.S., but most common in the East
An introduced species

Golden Shiner

Youngsters of this deep-bodied minnow start off silvery in color and become golden as they get older. It prefers to swim in quiet, weedy ponds, swamps, and streams, where it feeds on plankton, as well as water insects, mollusks, and algae. It is more likely that you will see these fish swimming in a big group than alone. Females lay their eggs in midsummer, and they stick to underwater plants.

Group: Carps and Minnows (Cyprinidae)
Size: Up to 12 ins
Distribution: Native to southern Canada, and eastern North America south to Texas, but introduced widely elsewhere

Northern Pike

This big fish is easy to recognize, with its long body and duckbill-shaped snout. Its jaws are lined with sharp, pointed teeth, with which it kills and eats many fish, as well as frogs, insects, mice, and even birds. Its body is greenish, with white or yellow spots. It swims in cold-water lakes, reservoirs, and weed-choked rivers. Left undisturbed, it can live as long as twenty-four years.

Group: Pikes (Esocidae) – **Size:** Up to 52 ins
Distribution: From Alaska throughout Canada, south to Missouri, and east to Pennsylvania and New York

Nine-spined Stickleback

This fish is slender, and, as its name suggests, has a row of spines on its back. It doesn't always have exactly nine spines. There can be between seven and twelve. The color is dull brown and blotchy above, and silvery on the belly. It may live in fresh or salt water, as long as it is cold. It prefers densely weeded, small ponds and streams.

Group: Sticklebacks (Gasterosteidae)
Size: Up to 2 ins
Distribution: Cold northern waters from Alaska north along coast to the Arctic Ocean and east to the Atlantic Coast, south through Canada to Indiana and New Jersey

Brown Bullhead

This is a type of catfish, so it has dark whiskers called "barbels" on its chin. It is a mottled chocolate-brown color with a yellowish or milk-white belly. As with all catfish, there is a small, rayless fin between its back and tail fins. The Brown Bullhead lives in muddy-bottomed ponds and lakes, and in slow streams and rivers. It uses its barbels to feel along the bottom muck for the insect larvae and mollusks that it eats. The female lays eggs in a scoop in the mud, which the male guards carefully.

Group: Catfishes (Ictaluridae) – **Size:** Up to 20 ins
Distribution: Southern Canada and eastern U.S. from Maine and the Great Lakes south to Florida and Mexico

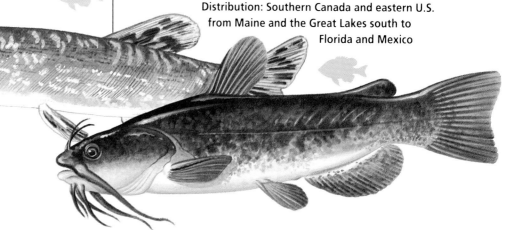

Northern Red-belly Dace

This small fish gets its name from the male which has a red belly in the breeding season. It has darker stripes along the greenish brown back and the belly is normally white. The scales are so small that the fish appears to have none. Its fins may have a yellow tinge, and its tail fin is forked. It is common around the swampy edges of ponds and lakes. It also lives in brooks, springs, and small streams which have gravel on the bottom. It feeds mainly on algae, but will also eat water insects.

Group: Carps and
Minnows (Cyprinidae)
Size: Up to 2½ ins
Distribution: Eastern U.S. from Wisconsin to Pennsylvania, south to Oklahoma and northern Alabama

Longear Sunfish

This brightly colored sunfish gets its name from its very long gill flap, which looks rather like an ear. It has blue-green, wavy lines on its stout body, and its belly is orange. Its back fin has spines which give it a jagged appearance. Look for this fish in lakes, reservoirs, and small, clear streams and rivers that have gravel and rock on the bottom. It feeds on water insects, crustaceans, and small fish. Males dig a nest in the gravel in the summer.

Group: Sunfishes (Centrarchidae) – Size: Up to 7 ins
Distribution: Southeastern Canada and eastern U.S. from both Dakotas east to the upper St. Lawrence River, and south to Florida and Texas

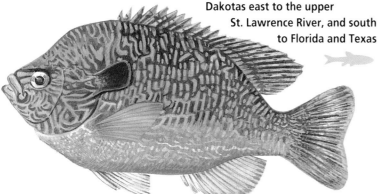

Fathead Minnow

This unusual-looking fish has a fat body and a head which looks flattened on the top. Its back fin is round and starts with a blunt-tipped half-ray. It has a dark olive back, with yellowish sides, and a white or cream belly. It lives in small streams, ditches, and boggy lakes with soft mud bottoms. In the mating season, the male's color darkens and he develops little knobs on his head, as shown. He uses these to push the female into his prepared nesting area. Fathead Minnows feed on algae and plankton.

Group: Carps and Minnows (Cyprinidae)
Size: Up to 3 ins
Distribution: Throughout central North America, west to the Appalachians, south into Mexico, introduced widely elsewhere

Pumpkinseed

This fish is also called the Common Sunfish. It has a fat body, which is tan or pale yellow, and its sides are covered in spots. There is also a scattering of tiny brown spots on the tail fin. A good identification guide is the black and orange tip of the gill cover. The back fin has stiff spines. You'll find the Pumpkinseed swimming in streams, ponds, and lakes with plenty of plants in them. It feeds on mollusks, insects, and fish. In the summer, males scoop out saucer-shaped nests in the bottom sand or gravel. Then females lay their eggs in them.

Group: Sunfishes (Centrarchidae) – Size: Up to 10 ins
Distribution: Canada, Atlantic Coast to Georgia, Great Lakes, and upper Mississippi River Valley south to southern Illinois, introduced widely elsewhere

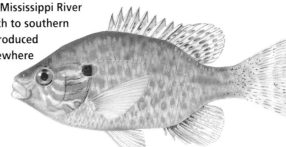

White Crappie

This fish is olive or yellowish in color with about ten faint bars running down the side of its body. The fins are very mottled in appearance. The White Crappie lives in streams, rivers, ponds, lakes, and reservoirs, where it feeds on water insects and plankton as a young fish. When it grows bigger, its diet changes to include fish, mollusks, and crustaceans. The male White Crappie builds a nest which it then guards, a habit which earned it the other name of Bachelor Fish. Many people like the taste of this fish, so it is very popular with anglers.

Group: Sunfishes (Centrarchidae) – **Size:** Up to 15 ins
Distribution: Southern Ontario west to Minnesota and Nebraska, south through the Mississippi basin in the Gulf states, north along Atlantic Coast to North Carolina

Largemouth Bass

This big fish gets its name from its big mouth which stretches back further than its eye. Its lower lip sticks out beyond the upper one. The Largemouth Bass is quite streamlined in shape, and is green with brown mottling. Its back fin has stiff spines. It lives in shallow, weedy, lakes, or in ponds or reservoirs, where it feeds mainly on insects, crayfish, frogs, and fish. The male builds a nest in the spring. When the young hatch out, they swim around in a group or "school," sometimes for as long as a month. The Largemouth Bass is one of the most popular game fish in the U.S. and is much sought by anglers.

Group: Sunfishes (Centrarchidae) – **Size:** Up to 36 ins
Distribution: Southern Ontario, west to Minnesota and Nebraska, south to the Gulf States, north along East Coast to North Carolina, introduced widely elsewhere

Bluegill

This fish may be blue, as shown here, but it can also be yellow or brown. There are several vertical bars on its sides. In the breeding season, males become bright orange on their bellies. The Bluegill has a small mouth and feeds on mollusks, worms, and water insects. It prefers to live in warm water habitats in weedy ponds, lakes, streams, and rivers. Males dig out shallow nests from the bottom sand during the summer. It is a favorite with fishermen.

Group: Sunfishes (Centrarchidae) – **Size:** Up to 10 ins
Distribution: Southeastern Canada and Atlantic Coast to Georgia, Great Lakes, and the upper Mississippi River Valley south to southern Illinois, introduced widely elsewhere

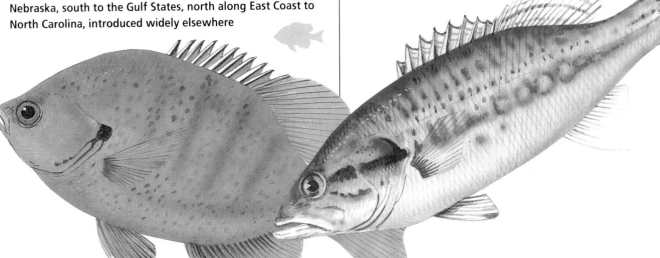

Explore a Pond

Exploring a pond is an easy way to find out more about freshwater life. Ponds are often teeming with little creatures, and the area around a pond is home to many others. Always take a friend and always tell an adult when you go exploring.

Pond dipping

Be very careful when walking by water.

Pond dipping is exactly what it sounds like: it involves dipping a net into a pond and seeing what you come up with. Spring is probably the best time to go pond dipping. Then, there is not too much weed, and there are lots of adult beetles, bugs, tadpoles, and newts.

If you want to see insect larvae such as dragonfly nymphs, go in the summer. They hatch in the spring, and by the summer they will have grown to their full size. Winter is best for creatures like fairy shrimps that live in temporary ponds. These types of ponds form in the fall and dry up in the spring.

1 **You will need a simple net with a mesh of about 2 mm**. If the mesh is too fine it will clog up. If it is too big, it won't catch anything small.

2 **Approach the water's edge quietly**. Any sudden noise or movement will send the pond creatures running for cover.

3 **Sweep your net across the surface of the pond** for surface-dwelling insects like beetles, pond skaters, water boatmen, and water mites.

4 **Scoop the net gently under pond weeds,** brushing the weed to knock off animals living and crawling on it such as tadpoles, damselfly nymphs, mayfly nymphs, or snails.

5 **Gently push the net along the surface of the bottom mud**. Be careful not to scoop too deep or it will fill up with muck. Here, you may find flatworms, caddisfly cases, worms, and shrimps.

6 **Put your finds into your container** with water from the pond.

7 **When you have finished examining them**, return them gently to the water.

Be careful which animals you put together: newts, beetle larvae, and dragonfly nymphs will eat other creatures such as tadpoles.

Examining very small animals

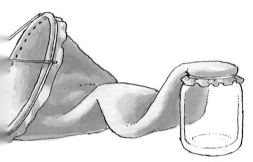

1 **Make a tapering net with a very fine mesh,** for example, from your mother's old nylon tights.
2 **Cut a hole in the end of the net** and tie the end around the mouth of a jelly jar.
3 **Walk round the pond trailing the net with its jar.** Keep it in the clear water above the bottom and away from weeds. Keep going for some minutes. The animals you catch will be washed down into the jar and will collect there.

4 **Lift the jar up and look at it.** It should be full of tiny moving specks, such as water fleas, and copepods. Use your magnifying glass to examine your catch.

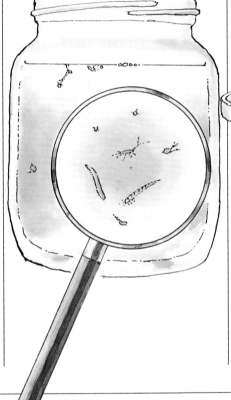

5 **If you have a microscope,** you can use it to look at your catch more closely. Use a dropper or the filler from an old fountain pen to transfer drops of water to a small shallow dish such as a jar lid. You will see them better if they can't move too far.
6 **Experiment with different colored backgrounds** and lights to see how best to show up the animals.

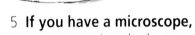

Mud grubbing

The mud at the bottom of a pond is a good place to look for worms and other animals that can put up with low oxygen levels and not much light. The best way to find these is to scoop up some mud and debris with your net or a bucket and sieve it, a little at a time (use an old kitchen sieve). Hold the sieve half in the water and shake it gently so that the mud floats away leaving the animals behind.

Mollusks

Modest Gyralus

This is a ramshorn snail with four rounded whorls. It is glossy and pale to dark brown in color. The snail is common and lives in many types of freshwater habitat, especially those with muddy bottoms. It lives on submerged water plants.

Group: Ramshorn snails (Planorbidae)
Size: Up to ¼ ins
Distribution: Found throughout Canada and the U.S.

Common Tadpole Snail

This snail is also called the Bladder Snail because its shell is like a sac or a bladder. It has a large opening on the left side, and only a tiny spire. It is yellow or brown in color. The Tadpole Snail lives in rivers, streams, ponds, and stagnant pools, and feeds mainly on plants. It needs to come to the water's surface to breathe.

Group: Bladder Snails (Physidae)
Size: ½–1 ins
Distribution: Found throughout Canada and the U.S.

American Ear Snail

This snail's shell has a huge opening in proportion to the rest of it. It is spiral in shape, with a short, narrow cone at the top, and is light greenish brown to yellowish brown. The last whorl before the opening has an earlike flare, which gives it its name. You'll find the Ear Snail in lakes, ponds, and slow-flowing streams among lily pads and reeds.

Group: Pond Snails (Lymnaeidae)
Size: Up to 1 ins
Distribution: Found throughout the eastern U.S., and introduced in the western U.S.

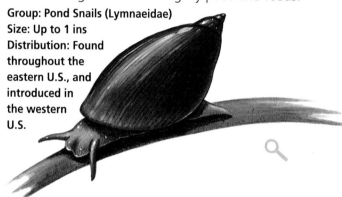

Three-whorled Ramshorn

Can you see how this small snail got its name? Its shell looks like a ram's horn and has three distinct whorls. Other species of ramshorn may have four or five whorls, but in all of them the shell is flat and coiled with no spire. Most are brown or chestnut-colored. The Ramshorn can be found in lakes, ponds, and slow streams, where there is plenty of plant life to feed on. It breathes air, so it must regularly come to the water surface to survive.

Group: Ramshorn Snails (Planorbidae)
Size: ½–1¼ ins
Distribution: Eastern U.S. diagonally northward to western Canada

Great Pond Snail

This snail has a tan-colored, almost transparent shell, with a sharp spire on the top. The first whorl is large, up to half the total height. It is also called the Stagnant Pond Snail as it can survive in stagnant waters. This is because it breathes air, and comes to the water surface at regular intervals. It can be found in large ponds and lakes, as well as in ditches and marshes. The Great Pond Snail is the biggest pond snail of all.

Group: Pond Snails (Lymnaeidae)
Size: 1½–2 ins
Distribution: Found throughout most of Canada and the U.S.

Florida Apple Snail

This large snail has a rounded, smooth shell that is olive-brown in color, with brown bands encircling it. It lives in still lakes and slow-flowing rivers, and can breathe oxygen in the water through gills. It feeds on plants. Females lay a huge mass of eggs on plant stems near the water. When they hatch, the young snails drop into the water.

Group: Apple Snails (Ampullariidae)
Size: 2–2½ ins
Distribution: Florida

Eastern Mystery Snail

This large snail has a glossy, olive-green shell with a round opening and a high spire. The whorls are rounded, and usually have brown bands. It gives birth to live, shelled young, rather than laying eggs. The Mystery Snail lives in ponds, lakes, and slow rivers, where it feeds on algae.

Group: River Snails (Viviparidae)
Size: 1½–2 ins
Distribution: Eastern U.S. and southeastern Canada

Valve Snail

The Valve Snail lives in deep lakes, usually on the mud among plant life. It has a broad, long shell with four rounded whorls, and is pale brown or brown in color. When the Valve Snail retreats into its shell, it seals the opening with a round, horny plate, the "operculum."

Group: Valve Snails (Valvatidae)
Size: Up to ¼ in
Distribution: Found throughout Canada and northeastern U.S., from Maine to Minnesota

Crustaceans

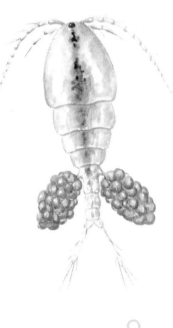

Copepods

These tiny creatures often occur in huge numbers. They live in any kind of still, freshwater habitat, from ponds to puddles. They use their legs or their antennae to swim along. Females have egg sacs which hang on either side of the body. They are an important food source for other freshwater animals.

Group: Copepods (Copepoda)
Size: Less than ¼ ins
Distribution: Found throughout North America

Fish Louse

This creature looks very different to the Water Louse. It has a flat body, a pair of suckers, and four pairs of legs. The Fish Louse is a parasite, which means that it lives off another creature. It clamps its suckers onto a fish's body or fins, and sucks up the blood. It is also a good swimmer, kicking along in the water with its legs. Look closely at the gills and skin of a pike or other freshly caught fish, and you may spot one.

Group: Fish Lice (Branchiura) – Size: Up to ½ ins
Distribution: Found throughout North America

Water Fleas

These tiny crustaceans have their whole body enclosed in a thick shell, leaving only the forked antennae free, which are used for swimming. They get their name from their jerky movement. Water Fleas vary in color, from greenish to brown or red. They can be found in great numbers, in many freshwater habitats, especially ponds. You will need a hand lens to see them clearly.

Group: Water Fleas (Cladocera)
Size: Less than ¼ ins
Distribution: Found throughout North America

Water Louse

This creature is a flattened version of the familiar land wood louse. The Water Louse can be found in almost all kinds of water, including puddles. It feeds on decaying plant and animal material. It is unable to swim, but crawls instead over mud and plants. In spring, some ponds teem with thousands of these little animals.

Group: Isopods (Isopoda)
Size: Up to ¾ ins
Distribution: Found throughout North America

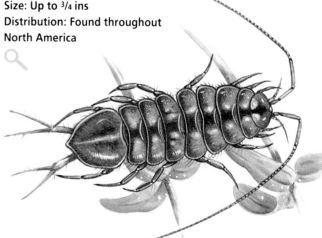

Chimney Crayfish

Crayfish look like miniature lobsters, but live in fresh water instead of in the sea. Many of them live in shallow streams, hidden under flat stones or in a shallow burrow. The Chimney Crayfish, however, is found in ponds, swamps, and meadows. Here, it digs a burrow up to three feet deep, with a chamber at the bottom filled with water. At the top of the burrow, it often builds a turret or chimney, several inches high.

Group:
Crayfishes (Astacidae)
Size: 2–5 ins
Distribution: Found throughout the U.S. east of the Rockies

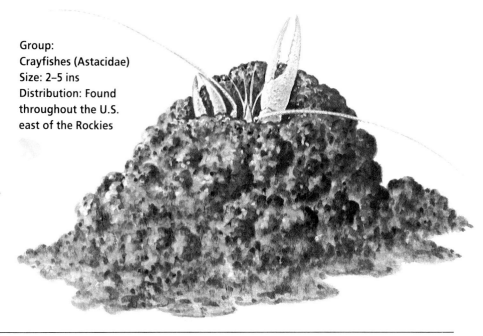

Fairy Shrimps

These little creatures live in small, temporary pools often in woodlands or old forests. They appear miraculously in spring when snow and ice melt or rain fills potholes. They swim along upside-down, using their eleven pairs of leaflike legs. This motion also brings them food in the water to filter. Females lay tough eggs which sink into the bottom mud. When the pool dries up, the eggs survive there and hatch out when a fresh supply of water fills the pool.

Group: Fairy Shrimps (Anostraca)
Size: Up to 1³/₄ ins
Distribution: Found throughout North America

Freshwater Prawns

These are also called Glass Shrimps because they are almost transparent. Most prawns and shrimps live in salt water, but freshwater prawns can survive in pools and ditches some way from the sea. They move slowly along the bottom, picking up bits of weed with their two pairs of pincers.

Group: Shrimps and Prawns (Natantia)
Size: 1–1¹/₄ ins
Distribution: Found throughout North America

Whirligig Beetle Larva

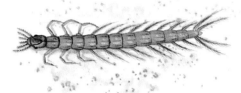

The young of the Whirligig Beetle looks rather like a centipede. However, it has only three pairs of legs. The rest of its legs are gills along the sides of its body. It hunts along the bottom of ponds and streams for other insect larvae and plants to eat. You can tell this larva from the similar alderfly larva because it does not have a pointed, hairy tail.

Group: Whirligig Beetles (Gyrinidae) – Size: up to 3/4 ins
Distribution: Found throughout North America

Dragonfly Nymph

The whirring wings and bright colors of adult dragonflies are a familiar sight around ponds and lakes. However, these beautiful insects start life as grotesque-looking, brown nymphs which live in the water. They are ferocious hunters with hinged jaws called a mask. When a tadpole or young fish is spotted, the nymph shoots out its mask and seizes the prey. It can swim by jet propulsion, sucking in and squirting water out from the end of its body.

Group: Dragonflies (Anisoptera)
Size: Up to 4 ins
Distribution: Found throughout
North America

Alderfly Larva

Adult Alderflies can be found hiding among plants by the water's edge. The young Alderfly, or larva, lives in the water at the muddy bottom of ponds, ditches, and streams. It has a long, brown body, with three pairs of legs and seven pairs of thin gills sticking out from it. The tail ends in a single hairy point. It is a hunter and catches its prey with strong pincer-like jaws.

Group: Alderflies
(Megaloptera) – Size: Up to 1 ins
Distribution: Found throughout North America

Mosquito Larva

Tiny, wriggling, worm-like mosquito larvae are a familiar sight in water butts, gutters, and any stagnant water. Look at them through a magnifying glass and you will see that each one has a bulbous head and breathes through a tube in its tail, which it sticks out of the water's surface. You will also find them hanging upside-down at the surface of water in small ponds, ditches, bogs, and marshes. Luckily they are a favorite fish food, which helps to keep their numbers down.

Group: Mosquitoes and Gnats (Culicidae)
Size: Up to 1/2 ins
Distribution: Found throughout
North America

Northern Caddisfly Larva

If you ever see a bundle of twigs walking across the bottom of your collecting bucket, it is probably a Caddisfly larva. The young of the Caddisfly looks like a caterpillar, if you can ever see it properly. The larva builds a special case to protect itself. The species shown will use anything available to make its case, from tiny sticks arranged in a crisscross pattern to snail shells and stones, all stuck together. Look out for cases on the bottom while pond-dipping.

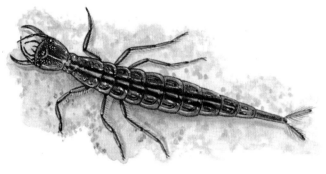

Group: Caddis or Sedgeflies (Tricoptera)
Size: Up to 1 ins
Distribution: Found throughout North America

Molanna Caddisfly Larva

This type of Caddisfly larva builds a different kind of case for its protection. It makes a neat tube, built out of sand, that itself sits on a sandy flat platform. It can be found on sandy bottoms of lakes or ponds, or slow-flowing streams.

Group: Caddis and Sedgeflies (Tricoptera)
Size: Up to 1 ins
Distribution: Found throughout North America

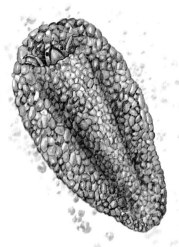

Diving Beetle Larva

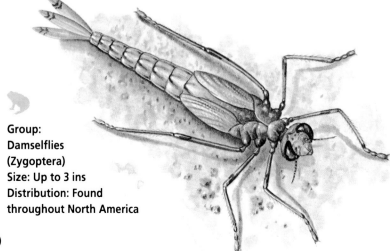

The adult Diving Beetle lives in the water and is a ferocious predator, but its larva is even worse! It can walk along the bottom mud, or swim, using its legs as oars. It will attack anything with its fierce, pincerlike jaws, from tadpoles and insects to fish. Watch out, as it can also give you a nasty nip. It sticks its tail out of the water to take in air through special tubes.

Group: Dytiscidae – Size: Up to 3 ins
Distribution: Found throughout North America

Damselfly Nymph

The adult Damselfly is like a small, slender version of the Dragonfly. Its nymph or young is also more slender than that of the dragonfly. It can easily be identified by the three leaf-like gills at its tail end. It cannot jet propel itself like the Dragonfly Nymph, but it hunts in a similar way, crawling about among the mud and weeds. Look for the empty shells of dragonfly and damselfly nymphs clinging to plants above the water's surface. These are left behind when the nymphs turn into flying adults.

Group: Damselflies (Zygoptera)
Size: Up to 3 ins
Distribution: Found throughout North America

Other Animals

Hydra

This tiny green or brown creature looks more like a plant than an animal. It has a slender body and about four to eight tentacles. When the hydra is hungry, it stretches these out to several times their normal length. The tentacles can sting and paralyze minute animals such as water fleas, which tumble into them. The Hydra then shortens its tentacles and pushes the food into its mouth. Hydras usually stay attached to water plants, but can let go and drift off somewhere else. Or they can move slowly along by turning somersaults. Look on page 71 to discover how you can find and keep these animals.

Group: Hydroids (Hydrozoa) – Size: Up to ½ ins
Distribution: Found throughout North America

Freshwater Sponge

Most sponges live in the sea, but a few small ones survive in fresh water. A sponge does not really look like an animal at all. Look out for a soft, spongy mass attached to sunken branches, stones, and wooden pilings. The shape varies according to where it is living. It may be flat, rounded, or have finger-like growths. It is usually greenish, but can be brown or gray in gloomy conditions. Sponges feed and breathe by drawing water into themselves through tiny tunnels. They can be found only in clean water in ponds, lakes, rivers, and canals.

Group: Sponges (Porifera) – Size: Up to 8 ins
Distribution: Found throughout North America

American Medicinal Leech

Leeches are types of worms that have muscular, stretchy bodies with a sucker at each end. The medicinal leech is greenish in color, with long stripes and bright orange spots in the middle. Its underside is rich orange in color and may have black spots. It sucks the blood of mammals, including humans, by first piercing the skin with its jaws. It then uses a special compound to let the blood flow freely. This leech is called "medicinal" because it used to be collected for medical purposes in the late nineteenth century.

Group: Leeches (Hirudinea)
Size: Up to 5 ins
Distribution: Found throughout North America

Horse Leech

This large leech lives in ponds or marshes, or wanders about in damp places searching for food. It has a stout, flattened body that is yellowish green on top and paler on the underside. The name is misleading because it rarely sucks blood from horses or other large animals. More often, it eats small animals such as earthworms or even small frogs. It swallows them whole, as it has weak jaws with few teeth.

Group: Leeches (Hirudinea)
Size: Up to 6 ins
Distribution: Found throughout North America

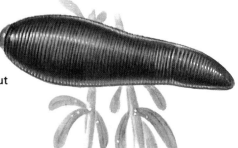

Leech (Helobdella)

This leech is similar in shape to the Snail Leech, but has a very soft body. It is pale gray, with a green, yellow, or brownish tinge. There is a tiny, hard scale on its back, but you will need to look very closely to see this! It lives in slow-moving rivers, lakes, and ponds, and feeds by sucking the body fluids of snails.

Group: Leeches (Hirudinea)
Size:
1/4–11/4 ins
Distribution:
Found throughout North America

Snail Leech

This leech has a ribby, flattened, leaf-shaped body that is dull green to brownish in color. It has several rows of yellowish spots and two broken, dark lines running along its back. As the name suggests, it feeds by sucking the body fluids of snails. The Snail Leech leads an inactive life, spending much of its time hidden under stones. It is common in still and running water.

Group: Leeches (Hirudinea)
Size: 1/2–11/2 ins
Distribution:
Found throughout North America

Leech (Theromyzon)

This leech has a soft, almost jellylike, flattened body. It is amber or gray in color, and large ones have rows of yellow stripes along the back. The underside is pale gray. It feeds by getting into the nasal passages and mouths of water birds, especially ducks, and sucking their blood. When satisfied with its feed, it blows up in size, and may take weeks to digest its meal. It does not swim.

Group: Leeches (Hirudinea)
Size: 1/2–11/4 ins
Distribution:
Found throughout North America

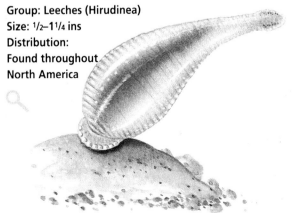

Fish Leech

As the name suggests, this leech feeds on fish. Its body is cylinder-shaped, with big suckers. It is greenish, yellowish, or brown, and may have white spots. When hungry, it "fishes" from a rock, waving about until a fish passes. Then it lets go and swims fast until it catches the fish and clamps tightly onto it. After sucking its blood, the leech lets go and hides among the plants. If you get one in your net, it will be easy to spot as it will loop rapidly about. These leeches are found wherever there are fish.

Group: Leeches (Hirudinea)
Size: 1–2 ins
Distribution:
Found throughout North America

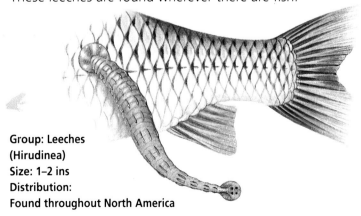

Keeping Tadpoles

Frogs and toads lay their eggs in ponds and streams or other wet places, mostly in spring or summer. Each egg is covered with a thick layer of jelly.

Frogs lay their eggs in large masses which are easy to spot. Most toads produce long strings of eggs and wind them around water plants. Newts and salamanders lay single eggs attached to leaves and stems.

Collecting spawn & tadpoles

The best place to collect spawn is from a friend's garden pond. Only collect from a "wild" pond if there is plenty there. Remember, it takes a great many tadpoles to produce just a few frogs because so many get eaten by fish, birds, and other predators.

1 **Collect about half a cup of spawn or a few dozen tadpoles** using a small aquarium net. Carry them home in a small bucket. Frog spawn is the easiest to find.

2 **Put them into a small aquarium tank** (see page 70). Use tap water that has stood outside for a day or two to get rid of the chlorine. Use pond water if it is not too muddy.

3 **Add a few rocks and some water plants.** Cover the tank with netting or the birds will get a free meal.

4 **Stand the tank in dappled shade.** Tadpoles like warm water, but may die if left in full sunshine.

5 **Your tadpoles will need feeding a few days after they hatch,** unless you have a large tank with lots of water plants. At this stage they are vegetarian. Add small pieces of boiled lettuce leaves and four to five pellets of rabbit food every three or four days.

6 **Change the water if it gets murky and fill it up as it evaporates.**

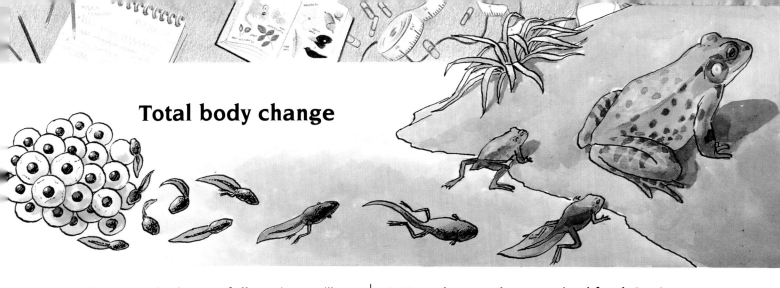

Total body change

1 **Watch your tadpoles carefully** and you will see them gradually change into adults. This change is called metamorphosis. Keep a diary of what happens:
   - Can you see the gills (salamanders only)?
   - Which comes first—hindlegs or front legs?
2 **When the hindlegs have appeared** (about five to six weeks after hatching), put some rocks or twigs in the tank for them to climb out on. They will soon need to breathe air.

3 **Now they need some animal food**. So give them small pieces of raw meat, about once a week. Remove uneaten food after a couple of days and change the water twice e week.
4 **When the front legs appear**, prop up the tank so that there is a shallow end. Or you can build some islands.
5 **The small froglets or toadlets should be released** by the edge of the pond from which you took the spawn. They are very difficult to feed now, so it is better to let them go. Carry them in a box with damp moss or grass. They will drown in a bucket of water.

Swamps & Marshes

Also called wetlands, these habitats are neither water nor really dry land. They are found mostly in lowland areas, especially where there is heavy rainfall or flooding. Wetlands that have trees growing in them are called swamps.

Bogs and marshes don't have trees. They develop in the shallows around lakes and ponds, and at the edges of quiet streams. There is little water movement, so plants such as reeds, sedges, and water-loving grasses can grow out into the water.

These places are ideal for animals that like to spend only part of their time in the water, such as snakes and alligators. There are also pools for permanent water creatures such as fish. This picture shows eight animals from this book; see how many you can identify.

American Alligator, Spectacled Cayman, Cottonmouth (snake), Green Frog, Stinkpot Turtle, Green Water Snake, plus a Green Anolis lizard (right) and a Red-eared Turtle (center on bank).

American Crocodile

You can tell a Crocodile from an American Alligator by its long, narrow snout, and also by its smaller size. The Crocodile is grayish green, with dark bands on its back and tail. Its skin is tough and knobbly. Its jaws are lined with sharp, peglike teeth, which it uses to seize prey such as crabs, fish, raccoons, and water birds. Even when its mouth is closed, some teeth are visible. This Crocodile lives in the bogs and mangrove swamps of Florida and the Florida Keys. Unfortunately, it is endangered by hunters and also by people destroying its natural habitat. You are much less likely to see a crocodile than an alligator.

Group: Crocodiles (Crocodylia) – Size: 7–15 ft
Distribution: On coast in extreme southern Florida and the Florida Keys

Spectacled Caiman

A relative of the Crocodile and Alligator, the Spectacled Caiman is smaller and colored light brown to light yellow, with bands on its back and tail. It gets the name "Spectacled" from the bony ridge between its eyes, which looks like the frame of a pair of spectacles. The Spectacled Caiman lives in swamps, rivers, canals, and ponds, and feeds on fish, small birds, mammals, and amphibians.

Group: Crocodiles (Crocodylia)
Size: 4–8½ ft
Distribution: Introduced to southern Florida from Central and South America

American Alligator

The alligator is the largest reptile in North America. Its thick, rough, knobbly skin is black, with creamy bands visible on its sides and tail. Its snout is broad and shovel-like, a feature which will help you to distinguish it from the American Crocodile. Its jaws are lined with sharp teeth. The Alligator lives in marshes, ponds, lakes, rivers, and swamps, where it spends its time sunbathing on the banks, or swimming under water with only the top of its head poking out. It feeds on different animals, from fish and frogs, to snakes and mammals. Alligators have often been hunted by people for their hides, but they are now a protected species in the wild.

Group: Crocodiles (Crocodylia) – Size: 6–19 ft
Distribution: Atlantic Coast southward from North Carolina to Florida Keys, west to south Texas; north to southeastern Oklahoma, and south to Arkansas

Reptiles

Green Water Snake

As its name suggests, this snake is usually greenish or brownish in color. There are faint black bars on its sides. Underneath, its belly is cream to brown and may have spots. The Green Water Snake lives in marshes, swamps, ditches, and canals, and is usually active by day, when it feeds on minnows and little fish.

Group: Colubrid Snakes (Colubridae) – Size: 30 –72 ins
Distribution: South Carolina coast through Florida, west to Louisiana and eastern Texas, north through eastern Arkansas to southern Illinois

Striped Crayfish Snake

This snake's skin is quite glossy and brown, with a wide yellow or orange stripe on the sides. There are three faint dark stripes running down its back, and its belly is plain and yellow to orange-brown in color. You will find this snake where there are water hyacinths growing, in swamps, ponds, and lakes. The Striped Crayfish Snake's name gives a big clue to its diet, which is purely crayfish. It catches a crayfish by encircling the creature with its body and holding it still while it swallows. Female Striped Crayfish Snakes give birth to live young between May and September.

Group: Colubrid Snakes (Colubridae)
Size: 12–24 ins
Distribution: Southern Georgia and the Florida peninsula

Cottonmouth

Group: Pit Vipers (Viperidae) – Size: 18–72 ins
Distribution: Eastern and southern U.S. from Virginia to upper Florida Keys, west to southern Illinois, southern Missouri, Oklahoma, and Texas
WARNING: Avoid—EXTREMELY POISONOUS!
This snake's bite can kill you

This is a large water snake with a flat-topped head that is wider than its neck. It can be olive, brown, or black in color, and may be plain, or have dark bands on its sides. Young Cottonmouths have strong patterns and bright yellow tails. This snake lives mainly in swamps, lakes, rivers, ditches, but can even be found in mountain streams. It swims along with its head out of the water, and sometimes can be seen basking on banks. It becomes most active at night, when it hunts frogs, fish, snakes, and birds. If you see this snake, keep well away as it gives a very poisonous bite.

Swamp Snake

The little Swamp Snake has a glossy, black coat and a red belly. The black color extends onto the edges of the smooth belly scales, giving an attractive pattern of triangles either side of the belly. As the name suggests, the Swamp Snake likes to live in swamps, especially where water hyacinths grow in large numbers. It can often be seen hiding among these plants, or at night after heavy rainfalls. The Swamp Snake feeds on leeches, small fish, frogs, and tadpoles. The females give birth to live young.

Group: Colubrid Snakes (Colubridae)
Size: 10–18 ins
Distribution: North Carolina coast, the Florida peninsula, southeastern Alabama, some in South Carolina

Plain-bellied Water Snake

This snake is so called because its belly is always plain red, orange, or yellow. The rest of the body is reddish brown, greenish, or gray, and is covered in rough scales. This water snake can be seen in river swamps and on the wooded edges of streams, ponds, and lakes. It feeds on frogs, fish, tadpoles, and crayfish. Look particularly in the early evening, as this is when it becomes most active. You may see it crossing roads on rainy nights.

Group: Colubrid Snakes (Colubridae) – Size: 30–60 ins
Distribution: Eastern and Southern U.S., from Delaware to northern Florida, to western Texas and southeastern New Mexico, in western Missouri, southern Illinois and Indiana, also in Michigan, Ohio, and eastern Iowa

Rainbow Snake

This snake is called the Rainbow Snake because of the rainbow of red, yellow, and black stripes on its body. Its belly is red and has a double row of black spots. The tail has a sharp spine on the end, which is harmless. The Rainbow Snake will burrow into sandy soil or under wet plants on the edge of streams and rivers. It comes out mainly at night to hunt for eels, its favorite food.

Group: Colubrid Snakes (Colubridae)
Size: 36–66 ins
Distribution: East Coast and southern U.S. from Maryland to central Florida, and west to Mississippi River

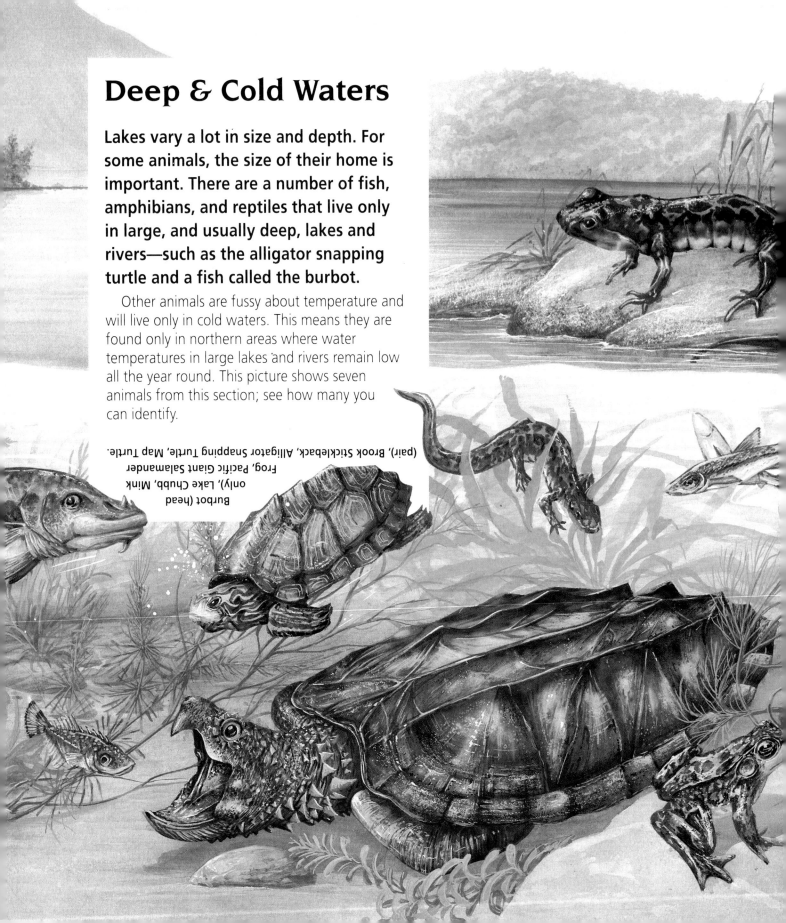

Deep & Cold Waters

Lakes vary a lot in size and depth. For some animals, the size of their home is important. There are a number of fish, amphibians, and reptiles that live only in large, and usually deep, lakes and rivers—such as the alligator snapping turtle and a fish called the burbot.

Other animals are fussy about temperature and will live only in cold waters. This means they are found only in northern areas where water temperatures in large lakes and rivers remain low all the year round. This picture shows seven animals from this section; see how many you can identify.

Burbot (head only), Lake Chubb, Mink Frog, Pacific Giant Salamander (pair), Brook Stickleback, Alligator Snapping Turtle, Map Turtle.

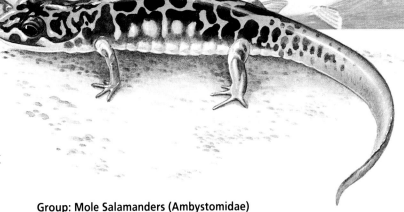

Pacific Giant Salamander

This fat-bodied, large salamander has smooth skin, with black marbling over a brown or purplish base. Its belly is light brown or yellowish white. The adult Pacific Giant Salamander can be found in cool, damp forests, or by clear, cold rivers and streams, under logs or rocks. Like most salamanders, the larvae have gills and live in the water, where they eat tadpoles and insects. Adults eat larger insects, mice, other salamanders, and small snakes.

Group: Mole Salamanders (Ambystomidae)
Size: 7–12 ins
Distribution: Southwest British Columbia and northwestern U.S. to California. Also Rocky Mountains in Idaho through to Montana

Mink Frog

Spotted Frog

This large frog is brown, with irregular, dark spots on its body. It has a light stripe on its upper jaw. Its belly is yellow, orange, or red, with darker mottling on the throat. The Spotted Frog lives near cold mountainous streams, rivers, and lakes, and is active during the day. Its voice sounds like a series of short croaks.

Group: True Frogs (Ranidae)
Size: 2–4 ins
Distribution: Northwest America from southeastern Alaska and British Columbia to western Montana, Wyoming, Idaho, and Oregon. Some in Utah, Nevada, and Washington

This frog is so called because it smells like a mink when rubbed or handled roughly. It is olive or brown, with dark patches on its sides and hind legs. Its belly is yellowish. This frog lives in cold northern ponds or lakes, especially where water lilies grow. During the night, when it is most active, the Mink Frog likes to sit on lily pads and croak in a low-pitched tone.

Group: True Frogs (Ranidae)
Size: 2–3 ins
Distribution: Eastern Canada and northeastern U.S.

Lake Chub

This little fish is a type of minnow. It is slender with a short head, large eyes, and a rounded snout. There are little barbels on either side of its mouth. Its body is brown above, and silvery or white underneath, with a darker line running along the side. The Lake Chub lives in cold lakes and small streams, where it feeds on insect larvae.

Group: Carps and Minnows (Cyprinidae)
Size: Up to 4 ins
Distribution: Found throughout Canada to northern U.S., south to Iowa, Colorado, and Idaho

Alligator Snapping Turtle

Map Turtle

This huge, fearsome-looking creature is the largest freshwater turtle in the world! It has a dark brown shell with three rows of knobs and a jagged edge. It also has a very long tail and a big head with a dangerous-looking, hooked beak. If it opens its mouth, it reveals a pink, wormlike structure. This acts like a fishing lure, to attract prey inside. These turtles live in deep rivers and large lakes. You may find difficult to spot one, as it spends most of its time sitting on the bottom of its watery home, with its mouth open, attracting prey. It will eat anything it can swallow, and can give a serious wound to humans.

Group: Snapping Turtles (Chelydridae)
Size: Over 24 ins
Distribution: Southern U.S. from Georgia and Florida coast to eastern Texas, north to Iowa and Indiana
BEWARE—this animal can give you a serious injury!

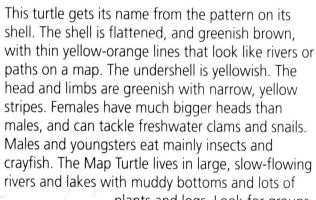

This turtle gets its name from the pattern on its shell. The shell is flattened, and greenish brown, with thin yellow-orange lines that look like rivers or paths on a map. The undershell is yellowish. The head and limbs are greenish with narrow, yellow stripes. Females have much bigger heads than males, and can tackle freshwater clams and snails. Males and youngsters eat mainly insects and crayfish. The Map Turtle lives in large, slow-flowing rivers and lakes with muddy bottoms and lots of plants and logs. Look for groups of Map Turtles sitting on top of each other on logs, basking in the sunshine.

Group: Pond, Marsh and Box Turtles (Emydidae) – Size: 4–12 ins
Distribution: Eastern central U.S. from Great Lakes south to Tennessee, Alabama, Arkansas, and Missouri.

Lake Whitefish

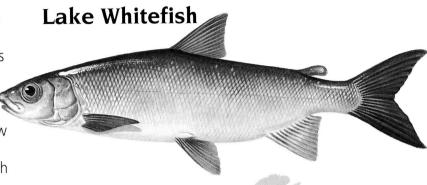

This slender, silvery green fish looks very much like the herring you find in the sea. However, it really belongs to the Salmon family and, like them, it has a small, fleshy, second (adipose) fin on its back. It has a small head and mouth and a forked tail. As its name suggests, the Lake Whitefish mostly lives in lakes, but it can also be found in wide, slow rivers. Large shoals are caught by commercial fishermen in nets. If they escape the nets, whitefish can live for over twenty years. Many do not even get the chance to hatch because the female scatters her eggs over the lake bottom where many are eaten by other fish and water birds.

Group: Trout and Salmon (Salmonidae)
Size: Up to 24 ins
Distribution: Found in most of Alaska, Canada, and northern U.S. including the Great Lakes

Lake Cisco

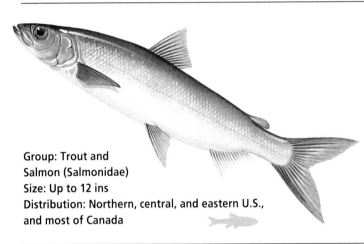

You may find it hard to tell the Lake Cisco from the Lake Whitefish as they are very similar. In fact, even the experts can't agree on how many different sorts of whitefish and ciscoes there are! The Lake Cisco has its lower jaw slightly longer than its upper, whereas the Lake Whitefish has it the other way round. It eats mainly plankton, crustaceans, and insect larvae. In turn, it is eaten by bigger fish and by people.

Group: Trout and Salmon (Salmonidae)
Size: Up to 12 ins
Distribution: Northern, central, and eastern U.S., and most of Canada

Lake Trout

Group: Trout and Salmon (Salmonidae)
Size: Up to 48 ins
Distribution: Alaska, Canada, and Great Lakes, eastward to New York and Maine

Most trout, like those on pages 62–63, live in fast flowing streams and rivers. The Lake Trout prefers deep, cold lakes and rivers. Long and slender, this large fish has a greenish brown body covered with white spots and a forked tail fin. Young fish feed on insects and crustaceans, and when they get older, they feed on fish such as ciscoes. Females lay their eggs in the fall, and they settle on the gravel bottom, where many get eaten by other fish.

Fish

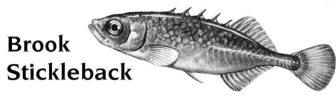

Emerald Shiner

This little fish is a type of minnow and gets its name from its shiny scales and the green stripe that runs along its side. The rest of its slender body is silvery-white. Its tail fin is forked and its back fin is transparent. The Emerald Shiner lives in lakes and large rivers, often in huge shoals. It feeds mainly on tiny floating plants and animals called plankton. It is often used as bait by anglers.

Group: Carps and Minnows (Cyprinidae)
Size: Up to 4 ins
Distribution: Found throughout Canada, south to Virginia and Texas

Longnose Sucker

The Longnose Sucker is so called because of the long, rounded snout that hangs over its lower lip. It is mostly a dark, mottled brown color, with a silvery-white underside. During the breeding season, males have a bright red stripe along the sides. This fish lives mostly in clear, cold waters of deep lakes and rivers, where it feeds on plant material and small animals. Females lay eggs in the late spring which stick to the bottom gravel.

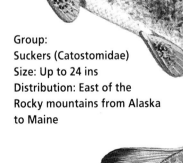

Group: Suckers (Catostomidae)
Size: Up to 24 ins
Distribution: East of the Rocky mountains from Alaska to Maine

Brook Stickleback

This little fish is easily recognized from the five to seven short, backward-pointing spines on its back. Like the more familiar Three-spined Stickleback (see page 14), it has no scales on its body. Its tail fin is rounded, and there is another spine underneath its body. The Brook Stickleback is brownish green, with lighter coloring underneath. It lives in the cool, clear waters of northern lakes and streams, and feeds on small crustaceans and insects. Males build a nest in April or May, which they guard fiercely—at other times you will see them in large shoals.

Group: Sticklebacks (Gasterosteidae) – Size: 3–4 ins
Distribution: Northeast U.S. from Nova Scotia and Maine on the East Coast, through New York west to Iowa and Montana; and most of Canada, north to Hudson Bay and southern Northwest Territories

Burbot

This is a strange-looking fish. Its body is long, almost eel-shaped, with fins that run along the back and underneath. There is a small barbel under its chin, which looks rather like a beard. It is usually mottled brown and yellowish, but the color is variable. The Burbot lives in large, cold rivers and lakes. It has a big appetite and feeds mainly at dusk and dawn on fish, crayfish, and other creatures. It breeds in the winter, under the ice if its lake is frozen.

Group: Cod Fishes (Gadidae) – Size: Up to 36 ins
Distribution: Alaska and Canada, south to Missouri and Ohio

Walleye

Group: Darters and Perches (Percidae)
Size: Up to 40 ins
Distribution: Northern U.S. and Canada, south through Great
Lakes and Mississippi River to
Louisiana, east to Alabama

The Walleye belongs to the same group of fish as
the perch and also has a forked tail and two back
fins. The first back fin has spines. However, it has a
much longer body. Its large eyes look glassy, hence
the name "Walleye." Its body is an olive-green
color, with several narrow, dark bands across the
back. Its tail has dark markings on it. The Walleye
lives in large streams, rivers, lakes, and reservoirs
that flow over sandy, or rocky bottoms. It
feeds mainly on other fish, and breeds
in the spring in shallower waters.

Yellow Perch

This fish can be recognized by its golden-yellow
body, which has six to eight dark bars running
down it along the side. Its tigerlike coloration has
given it another name, the Tiger Trout. Scales are
easy to see on its body. The first of its two back
fins has spines, as does the fin underneath its body.
Its tail fin is forked. The Yellow Perch lives in large,
clear streams, lakes, and reservoirs which have
plant life. It swims around in groups, feeding on
insects, snails, and other fish. Females lay eggs in
a ribbonlike string embedded in a jellylike
substance. These strings sometimes contain as
many as 80,000 eggs!

Group: Darters and Perches (Percidae)
Size: Up to 15 ins
Distribution: Eastern Canada, south along the Atlantic Coast to
Florida, and through the upper Mississippi River south to
Kansas. Widely
introduced
elsewhere

White Bass

This fish is plump-bodied and silvery in color, with
six to ten dark stripes running along its sides from
head to tail. The scales are medium in size. It has
two fins on its back, and the first has spines. There
are also spines on the fin underneath its body and
one on the gill cover. Its lower jaw protrudes
beyond its upper jaw. The White Bass lives in large,
clear rivers that flow over gravel or rubble. Females
lay eggs in late spring or early summer, which settle
on the gravel bottoms. White Bass swim in big
groups, and feed on small fish.

Group: Bass (Moronidae) – **Size:** Up to 15 ins
Distribution: Southern Great Lakes and Mississippi River valley
from Minnesota to Texas, and across the Gulf states to Florida

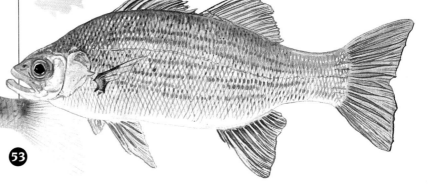

What Can I Catch?

Make a minnow trap

An easy way to study fish is to catch them and take them home. This trap should catch minnows, darters, or catfish.

1 **Make two cone shapes** out of wire mesh.
2 **Fix the wide open end of each one** onto a circle of stout wire.
3 **Push the narrow ends back inside** to make two smaller cones pointing inward.

4 **Cut the ends off the small cones to leave a small hole** big enough for small fish to swim through (about 2–3 inches across).
5 **Fix the two halves together** with a wire hinge and bulldog clip so it can be easily opened.
6 **Bait the trap** with dry dog or cat food or scraps of meat and fish. Put the bait in a fine mesh bag and hang it inside the trap.
7 **Set the trap some way out in a pond, lake or stream**. Let it drift downstream from a promontory or ask an adult to help if you need to use a boat.
8 **Anchor it with a rope** to a bush or tree and leave it for several hours or overnight.
9 **Don't forget to check it next day** or the fish will die.

Adopt a site

If you have a favorite pond, stream or swamp near you, visit it regularly (perhaps once a month) and get to know it really well. You will be able to find out which animals live there, and which are just passing through. What happens in winter and in spring? Make a plan of the site, name the different parts and mark on where the animals are found. You could repeat this at different times of the year.

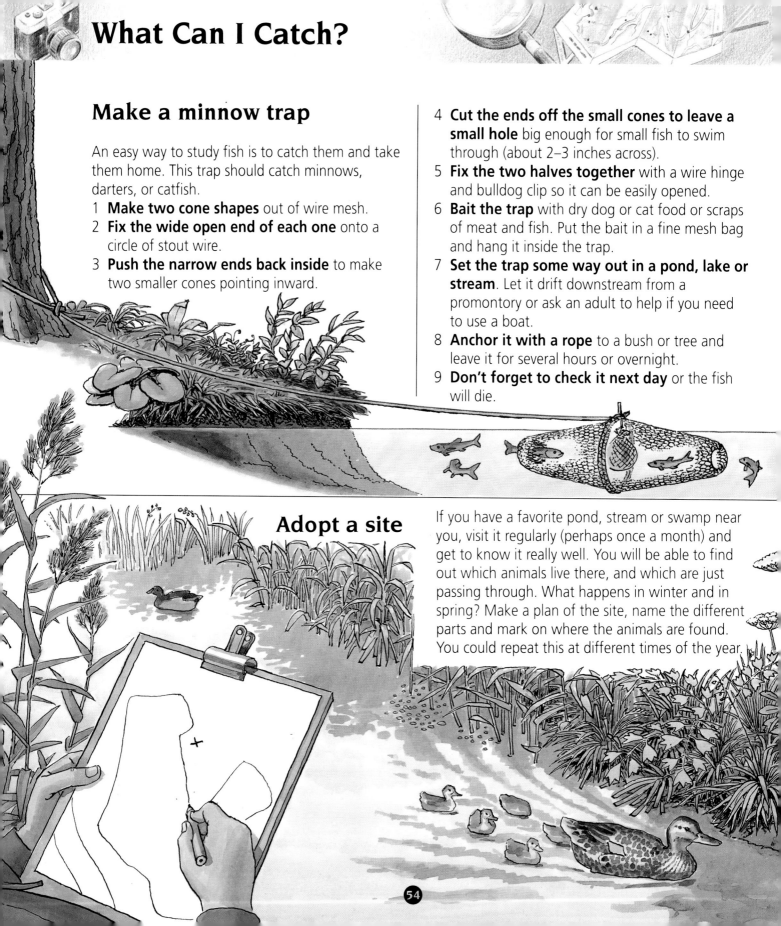

Stream bottom sampler

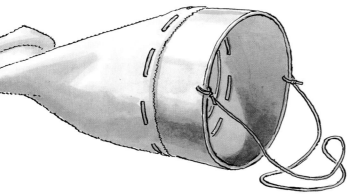

1 **Remove the bottom from a large biscuit tin** or a similar container such as a Tupperware box.
2 **Fix a conical net over one end.**
3 **Hold or weigh the box down on the stream bed** with the open mouth facing upstream so that the current flows through the net.
4 **Use a stick to stir up the bottom sediment and turn over stones** in front of the box. The dislodged animals will be swept through the box and into the net.

A trap placed on the bottom of a fast-flowing stream is ideal for catching the creatures that live there, as they get swept into it by the current. This is good for catching water skaters, freshwater shrimps, mayfly larvae, and even small fish.

Pollution Watch

Is your local river or stream polluted? Here are some of the signs you should look for.

- **Water changes color or is cloudy** (except after heavy rain)
- **Metal garbage**: like shopping carts, old tires, baby carriages
- **Sewage**: disposable diapers, toilet paper, etc.
- **Thick, green algae** covering the surface—this is a sign of nitrogen pollution from farms
- **Bad smell** rising from the water
- **White or brown foam** on the surface—probably from a factory spill
- **A line of grease** on rocks and boulders above the usual water level
- **Oil or gas spills**—this shows up as rainbow-colored circles floating on the surface
- **Dead fish** floating in the river or washed up in the shallows

If there is a marked change in the river, alert an adult as soon as you can and get them to inform the local health officials.

Fast Streams & Rivers

Many rivers begin life far up in the hills or mountains. Sweeping away mud or sand, the swift current runs over boulders, stones, and gravel. On steep downhill sections, there will be waterfalls and white-water rapids. The rushing water provides plenty of oxygen, but the animals living here must be tough enough to prevent themselves being washed away.

Many insect larvae can live here. Look for the flattened nymphs of mayflies and stoneflies clinging onto stones. Some caddisfly larvae anchor themselves to the bottom in silken nets. Wet boulders are favored by blackfly larvae which have hooks to cling on with. Look for the dusky salamander and the two-lined salamander which feed on the insect larvae hiding beneath stones.

Few snails are found in fast waters, but pearl mussels fix themselves to the stream bed and rely on a current to bring them food. Crayfish cling to the sides of rocks with their strong legs and claws.

Small fish like sculpins are flattened on their lower side and are streamlined. They lie with their head pointing upstream so that the water flows over them. Trout and salmon are also streamlined and are powerful swimmers. This picture shows thirteen animals from this book; see how many you can identify.

Bass, Black Fly nymph, Caddisfly nymph, Longnose Dace, Tailed Frog, Mayfly nymph, River Pearl Mussels, Atlantic Salmon, Red Salamander, Two-lined Salamander, Freshwater Shrimps, Stonefly nymph, Rainbow Trout.

Amphibians

Tailed Frog

This frog gets its name from the pear-shaped "tail," which is only present in males. It is colored olive, gray, or black, with lots of dark spots and some small warts on its back. Usually, a dark stripe runs from the snout through the eye, and the snout itself has a yellowish triangle. The Tailed Frog lives in cold, clear, fast-flowing mountain streams, or in nearby damp forests. Its tadpoles have sucking mouthparts that help them to cling onto rocks and logs in strong currents. Unlike most other frogs, this one does not make any sound.

Group: Tailed Frogs (Ascaphidae) – Size: 1–2 ins
Distribution: From south British Columbia along coast to northwestern California; some scattered eastward to northwestern Montana and Idaho

Cascades Frog

You will recognize this frog by the inky-black spots on its brown to olive-brown back. It has a dark patch covering the eye and ear drum, a light jaw stripe, and a yellow belly. This mountain frog lives near small streams, ponds, and lakes. You may see it in the water, or hiding among the grass. When frightened, it swims away rather than diving to the bottom. The Cascades Frog is active by day, and makes a sound like the twang of a loose banjo string.

Group: True Frogs (Ranidae) – Size: 1¾–2¼ ins
Distribution: Olympic mountains of Washington, and Cascade mountains of Washington, Oregon, and northern California

Hellbender

This frightening-looking creature is, in fact, quite harmless. Also called the Devil Dog, it is a giant salamander that never leaves the water. It looks slightly squashed and has a sideways flattened tail. It seems to be too small for its skin, which is loose and wrinkled. Most Hellbenders are gray or olive-brown with a paler belly. Some have spots and blotches. It lives in clear streams and rivers. You can find it by searching under rocks, where it likes to hide. It feeds on crayfish, snails, and worms.

Group: Giant Salamanders (Cryptobranchidae) – Size: 12–30 ins
Distribution: Eastern U.S. from southwestern New York to northern Alabama and Georgia, also some in Missouri and Susquehanna River

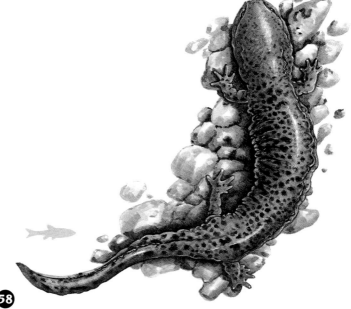

Foothill Yellow-legged Frog

This frog is gray, brown, olive, or reddish in color, and usually has a mottled pattern. It gets its name from the yellow underside of its legs. It likes to live in gravel or sandy-bottomed streams that have sunny banks for basking and woodlands nearby to explore. It is active mainly by day, and makes a grating sound, but this is rarely heard.

Group: True Frogs (Ranidae)
Size: 1–3 ins
Distribution: Western Oregon to southern California, except central valley of California

Dusky Salamander

This salamander is tan or dark brown, and may have mottling on its body. Young types have several pairs of round, yellowish or reddish spots running down the back. The tail is triangular in shape. You will find the Dusky Salamander near rocky woodland creeks and streams, where the rocks provide shelter. It feeds on insect larvae, bugs, and earthworms.

Group: Lungless salamanders (Plethodontidae)
Size: 2½–5½ ins
Distribution: Found in a wide band from southeastern Canada, running southwest to Louisiana

Red Salamander

Group: Lungless Salamanders (Plethodontidae)
Size: 2½–6 ins
Distribution: Southern New Brunswick and southeastern Quebec southwest to Louisiana

This stocky salamander is easy to spot with its short tail, bright red coloring and black spots. Older ones may become purplish brown. If you get close, you will see it has yellow eyes. The smaller Mud Salamander (see page 73) has brown ones. The Red Salamander lives in and near springs, streams, cool brooks, and nearby woodland. It may wander some distance from the water, and may be seen hiding in woodland leaf litter, or under rocks, or stones.

Amphibians

Two-lined Salamander

It's easy to see how this salamander gets its name. It has two dark lines bordering a broad, light stripe running right down its back. The broad stripe may be yellow, brownish, greenish, or orange-bronze. The Two-lined Salamander lives by rock-bottomed brooks and springs, and in river swamps and damp forests. It hides under rocks and leaf litter, and if alarmed will run or swim away rapidly.

Group: Lungless Salamanders (Plethodontidae)
Size: 2–5 ins
Distribution: Southeastern Canada and eastern U.S.

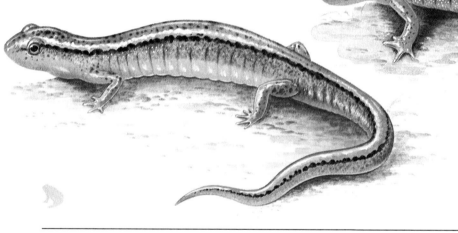

Olympic Salamander

This small salamander is plain brown or mottled olive in color. It has a small head with bulging eyes, a slender body, and a short tail. Underneath, its belly is yellow-green or yellow-orange, with black flecks. The Olympic Salamander lives in cold, fast-moving streams and springs near forests. When on land, it hides under stones. Its favorite foods include insects and spiders.

Group: Mole Salamanders (Ambystomidae)
Size: 3–5 ins
Distribution: West Coast U.S. from the Olympic Peninsula (Washington) to California

Long-tailed Salamander

One of the first things you will notice about this salamander is its extremely long tail, which is slender and much longer than its body. The body is yellow to bright orange-red, scattered with black spots that turn into bars on the tail. The Long-tailed Salamander likes to live by streams and springs near woodlands. On warm, rainy nights it may be seen wandering around on the forest floor in search of food.

Group: Lungless Salamanders (Plethodontidae)
Size: 4–8 ins
Distribution: Eastern U.S. from New York to Florida; west to the Mississippi River, Kansas and Oklahoma

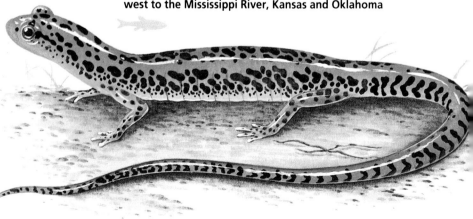

Black-bellied Salamander

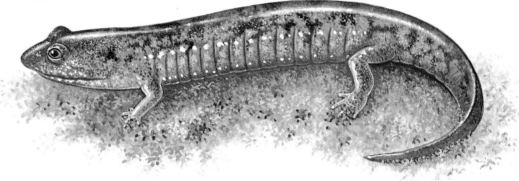

Group: Lungless Salamanders (Plethodontidae)
Size: 3½–8 ins
Distribution: southern West Virginia through the mountains to northeastern Georgia, some in South Carolina and Georgia Piedmont

This fat-bodied salamander is black with greenish blotches on its back, and black-bellied, just as its name says. It likes fast-flowing mountain streams that have boulders in them for sunbathing on. It rarely ventures far from the water's edge. Mainly active at night, the Black-bellied Salamander feeds on insects, snails, and small salamanders.

Spring Salamander

Group: Lungless Salamanders (Plethodontidae)
Size: 4–9 ins
Distribution: Southern Quebec and southern Maine to northern Georgia, Alabama, and Mississippi

This large salamander may be salmon, brownish pink, or reddish. Hazy, darker markings on top of this color give it a pattern. A light bar runs from its eye to its nostril. Called the Spring Salamander because it prefers cool springs, you may also spot this salamander in mountain streams, or beneath logs or stones in surrounding woodlands.

Atlantic Salmon

You can recognize all salmon and trout by their streamlined shape and small, fleshy second back fin. Most salmon have a slightly forked tail, whereas in trout, it is almost straight. Adult Atlantic Salmon are silvery with a few scattered spots (below). During the breeding season, the male develops red spots and a reddish belly. His lower jaw grows forward into a hooked shape. The Atlantic Salmon leads a complex life. It spends most of its life in the sea, but returns to the stream in which it hatched in order to breed. It will struggle up waterfalls and rapids to get there. After spawning, some die, but many return to the sea and will later spawn again. The young, or parr, are quite dark and have blotchy sides. At two to six years old, they start to move downstream, become silvery, and are called smolts.

Group: Trout and Salmon (Salmonidae)
Size: 48–54 ins
Distribution: North Atlantic and coastal waters from Arctic Circle south to Delaware River, Lake Ontario, plus some landlocked types in New England states

Rainbow Trout

This fish gets its name from its beautiful rainbow of blue, green, and pink that covers its body (above). Its body, tail, and back fin are also heavily speckled with small, dense black spots. Rainbow Trout live mostly in clear, upland stream pools and in lakes. However, some Rainbow Trout along the Pacific Coast migrate down to the sea when they are young. These ones are called Steelheads. They eventually return from the sea to breed in their own rivers and streams. The Rainbow Trout is one of the fish most sought after by anglers.

Group: Trout and Salmon (Salmonidae)
Size: 40–45 ins
Distribution: Native to Pacific Coast from Alaska south to northern Mexico, introduced into Canada, and eastern U.S.

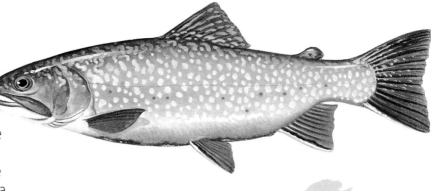

Brook Trout

You can recognize the beautiful Brook Trout or Brook Charr by its olive-green body, which is covered with wormlike patterns. Red spots surrounded by blue are scattered on its sides. The fins underneath its body are pink, edged with white, and males often have orange bellies in the breeding season, which is in the fall. Its tail fin is a squarish shape, which gives it the other name of Squaretail. The Brook Trout prefers to live in cold, clear streams.

Group: Trout and Salmon (Salmonidae)
Size: Up to 20 ins
Distribution: Eastern Canada and eastern U.S.,
Great Lakes region south through Appalachians to Georgia

Coho Salmon

Another name for this fish is the Hooknose—can you see why? Only the breeding males develop the hooked jaws and the red color along the sides. Its usual color is silvery with scattered, small black spots on its back and on the top of the tail. Like the Atlantic Salmon, the young fish migrate downriver to live in the sea, but they do not go far offshore. They return upriver to spawn when they are three or four years old, swimming tirelessly until they get to the small headstreams. After spawning they all die, and for a while, the streams are choked with their bodies.

Group: Trout and Salmon (Salmonidae)
Size: Up to 36 ins
Distribution: Pacific Ocean and rivers
joining it, from Alaska to southern
California. Introduced to the Great Lakes

Brown Trout

This trout is greeny-brown with silvery sides and is speckled with large, dark spots. It also has scattered red spots faintly surrounded by blue. Unlike the Rainbow Trout, it has an unspotted tail. The Brown Trout lives in moderately fast rivers and streams, often hiding in quiet pools. However, like the Rainbow Trout, some Brown Trout migrate to the sea, when they are called Sea Trout. Brown Trout become most active at dusk, when they will "rise" to snap up surface insects. They also feed on mollusks, crustaceans,   and even smaller Brown Trout.

Group: Trout and Salmon (Salmonidae)
Size: Up to 40 ins
Distribution: Introduced to North America
from Europe and Asia; now widely
stocked across much of Canada and the U.S.

Fish

Central Stoneroller

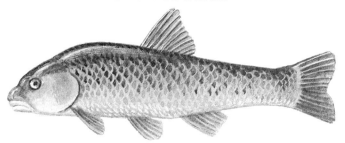

This fish is slender with a slightly hunchback appearance. It is brownish on top and silvery-white underneath. Dark scales are easy to spot, scattered over its back and sides. There is a single fin on its back, and its snout overhangs its mouth, giving the fish a bad-tempered appearance. The Stoneroller gets its name from the way in which it rolls stones and gravel on the bottom of the streams where it lives. It does this to find food such as insect larvae and mollusks, and also to dig a nest in the spring.

Group: Carps and Minnows (Cyprinidae)
Size: Up to 8 ins
Distribution: Southern Ontario, south through most of eastern U.S. except for southern Atlantic coastal states, and west to the Rockies

Brassy Minnow

This little fish gets its name from its brassy yellow sides, which glitter in the sunshine. The Brassy Minnow is slender, with a single fin on its back and a forked tail fin. Its medium-sized scales are easy to see. It lives in small, weedy creeks, where it creeps along the bottom searching for food in the slime. It lives in big groups, or "schools," and breeds in the spring.

Group: Carps and Minnows (Cyprinidae)
Size: Up to 4 ins
Distribution: From Kansas northward to Canada, east to New York, west to Montana and southern Canada

Prickly Sculpin

This sculpin is dark-colored on top and lighter underneath. There are two fins on its back which merge into one, and the first one is spiny. Like other sculpins, this one has very big, wavy fins on its sides (pectoral fins). It lives in the quiet areas of coastal streams, and spends most of its time hiding, only coming out to catch small animals to eat.

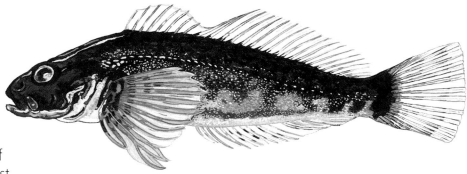

Group: Sculpins (Cottidae)
Size: 3–4 ins
Distribution: Pacific Coast of North America from California north to Alaska

Mottled Sculpin

Like most sculpins, this strange-looking fish has a large, flattened head and thick lips. It has a rather lumpy appearance. As its name suggests, it is mottled, with brown, black, and gray markings. Its eyes are on top of its head and its body has no scales. There are two large fins on its back, and the first one is black and spiny. The Mottled Sculpin lives in clear, cold streams, rivers, and lakes over rocky and gravel bottoms.

Group: Sculpins (Cottidae)
Size: Up to 4 ins
Distribution: All of southern Canada and most of central and western U.S.

Longnose Dace

Named after the snout which extends well past its mouth, the Longnose Dace is a long, slender fish. Its body is mottled with black and brown markings, and its underside is cream-colored. There is a single fin on its back, and its tail fin is slightly forked. This fish lives in schools, in swift-flowing streams and rivers that have gravelly bottoms. It feeds on small insects.

Group: Carps and Minnows (Cyprinidae)
Size: Up to 7 ins
Distribution: Canada, northern U.S. south in Appalachians to Georgia, south in Rockies to the Rio Grande, Texas, and New Mexico

Slimy Sculpin

This little fish has a slender, scaleless body which is greenish on top and creamy-yellow underneath. Its fins are pale gray in color. Two long, back fins almost merge into one, and the first one has seven weak spines. The Slimy Sculpin lives in deep, cool lakes and fast-flowing rivers and streams that have gravel and rocks on the bottom. In late spring, males dig a hollow under rocks and stones, then they entice females to lay eggs there with elaborate displays. The female goes into the nest and lays her sticky eggs on the "roof."

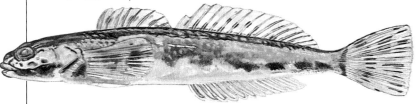

Group: Sculpins (Cottidae)
Size: Up to 4 ins
Distribution: Alaska and Canada, from the Great Lakes south to Virginia, and Maine to Oregon

Fish

Creek Chub

The Creek Chub's body is long, slender, and bronze, brown, or olive in color. A broad, dark band runs along its side from the snout to the beginning of the tail where it ends as a small spot. There is one fin on the back and the tail fin is forked. It has a large mouth and a pointed snout which extends over the lower jaw. The Creek Chub lives in small, fast brooks and streams that flow over gravel, sand, or rocks. It feeds on insects, worms, and small fish.

Group: Carps and Minnows (Cyprinidae)
Size: Up to 12 ins
Distribution: Found throughout most of U.S. east of the Rockies

Smallmouth Bass

This fish has a stout, olive-brown to bronze body. It is so called because its mouth is not as big as other types of bass. Its back fin is in two parts, the first jagged and spiny, the second soft-rayed. The Smallmouth Bass prefers to live in clear, rocky-bottomed or gravelly, mountainous lakes and streams, where it feeds on insect larvae, crayfish, and fish. In the spring or summer, the male digs out a nest in the bottom gravel and then entices a female to lay eggs in it. It is a favorite with anglers.

Group: Sunfishes (Centrarchidae)
Size: Up to 24 ins
Distribution: Southeastern Canada and eastern U.S. from Quebec, Ontario, and Great Lakes, west to South Dakota and Iowa, south to Oklahoma, Arkansas, and northern Alabama

Brook Silverside

A slender body that looks shiny and transparent, and a small, pointed head are the trademarks of this fish. It is yellowish in color, with a silvery band along its sides. Underneath, it is silvery-white. There are two fins on its back and the tail fin is forked. The Brook Silverside lives in calm areas of streams, rivers, and lakes that are full of plant growth. It feeds on insects from the water surface. In fact, it spends most of its time near the water surface, and often leaps out of the water.

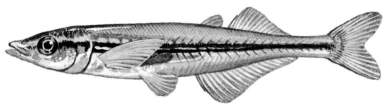

Group: Silversides (Atherinidae)
Size: Up to 4 ins
Distribution: Found in most eastern states and the Mississippi Valley to Minnesota

Sailfin Molly

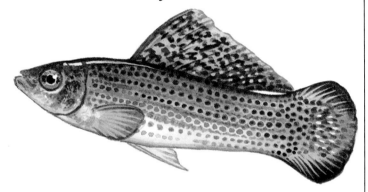

One look at this beautiful fish will tell you the reason for its name. It has a huge fin on its back that looks a bit like a sail on a boat. The body is not typically fish-shaped, as it is rather wide for its entire length. It is olive-brown in color, with rows of colored spots. The tail fin is rounded, with a black margin. Males have much brighter colors than females. The Sailfin Molly lives in fresh and brackish water areas of estuaries, swamps, and streams. Females give birth to live young, instead of laying eggs. They are often kept in aquariums.

Group: Live Bearers (Poeciliidae)
Size: Up to 3 ins
Distribution: From South Carolina to Mexico in coastal regions

Rainbow Darter

The blue, green, orange, and red colors on this fish's body earned it the name Rainbow Darter. It has a fairly stout body with a large head and a blunt snout. There are two fins on its back. The first is spiny and the second has soft rays and is higher than the first. Its tail fin is long and square in shape. The Rainbow Darter swims in clear, cool streams and small rivers that flow over gravel. It feeds on small water insects, snails, and small crayfish.

Group: Darters and Perches (Percidae)
Size: Up to 3 ins
Distribution: Southern Canada and eastern U.S. from Ontario and New York west to southern Minnesota, south to northern Alabama and Arkansas, southwestern Mississippi and eastern Louisiana

Other Animals

Black Fly Larva

Swarms of tiny, biting Black Flies can be enough to spoil a summer picnic. The young of these annoying flies cover rocks and plants in small, fast-flowing streams. Look carefully at the tops of rocks and around waterfalls and you are bound to see them. They fix themselves to the rock by a pad of silk into which they lock their tail end. They strain food from the water with two foldable fans of bristles.

Group: Black Flies (Simuliidae)
Size: Up to ¼ ins
Distribution: Found almost everywhere

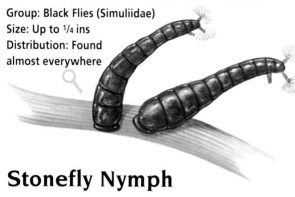

Stonefly Nymph

Like many other waterside insects, the Stonefly hatches out as a nymph that lives under water. Also known as creepers, the nymphs have strong legs for clinging beneath stones in the fast currents of the streams in which most species live. A pair of long antennae and two thin tails called "cerci" will help you to distinguish them from Mayfly Nymphs (see opposite) with which they are often found. Although usually plain brown, some species have beautiful patterns.

Group: Stoneflies (Plecoptera) – Size: Up to 1 ins
Distribution: Found almost everywhere

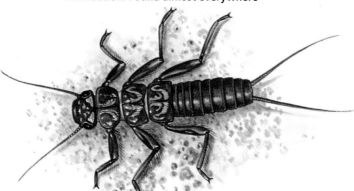

Caddisfly Larva

Most caddisfly larvae build themselves a case of twigs or stones to live in. Find out how to watch them doing it on pages 38–39. The species shown here acts rather like a spider. Instead of a case, it spins a funnel-shaped net of silk under a stone. Small animals are swept into it and eaten. If you look down into a clear stream, you might see crescent-shaped objects on the stream bed. These are the entrances to the silken nets. They live in fast-running water.

Group: Caddisflies (Hydropsyche) – Size: Up to ¾ ins long
Distribution: Found almost everywhere

Mayfly Nymph

There are many different kinds of mayflies, but all of them live near water. The nymph or young live under water mostly in streams, rivers and large lakes. Although different species vary in size and shape, all mayfly nymphs have a row of feathery or plate-like gills along the sides of the body, and three thin tails. Most are brownish in color. The one shown here has short, sturdy legs and burrows into gravelly mud. Others cling beneath stones along with stonefly nymphs.

Group: Mayflies (Ephemeroptera) – Size: Up to1 ins
Distribution: Found almost everywhere

Freshwater Shrimp

Freshwater Shrimps resemble the sand hoppers and beach fleas that you find on the seashore. They are flattened sideways and curved into an arc shape. There are several different species which vary in color from grey to greenish or orange-brown. You will not find them in stagnant ponds, but they are common in almost any clean, running water and in large lakes. Fish love to eat them, so they hide under stones and among plants. They eat decaying plants and animals.

Group: Amphipods (Amphipoda)
Size: Up to 1 ins
Distribution: Eastern U.S.

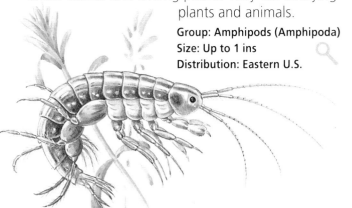

Eastern Crayfish

There are many different species of crayfish found throughout North America. They are easy to recognize because they look like miniature lobsters. However, it is very difficult to tell one species from another. The Eastern Crayfish is brown in color and has a smooth shell. Like other crayfish, it has ten legs, two of which have sharp pincers on the end for pulling its food apart. Crayfish make a tasty meal for fish, birds, turtles, otters—and humans!

Group: Crayfishes (Astacidae)
Size: 2–5 ins
Distribution: Eastern U.S. from Tennessee and the Carolinas to Maine

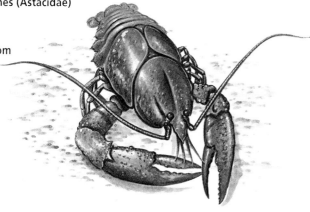

River Pearl Mussel

This mussel lives on sandy bottoms of cool, clear streams and small rivers. Its thick, heavy shell is yellow-brown when young, then the color changes to dark brown to black when it gets older. Inside, its shell is pearly white. Sometimes the River Pearl Mussel forms small pearls. Like some other mussels, the newly-hatched young attach themselves to fish and feed off them for a few weeks. They grow very slowly and can live to be ninety years old!

Group: Pearly Mussels (Margaritiferidae)
Size: 3–6 ins
Distribution: Eastern River Pearl Mussel on Atlantic Coast from Labrador to Pennsylvania; Western River Pearl Mussel on Pacific Coast from California to Alaska

Keep Them at Home

It's easy to make your own temporary aquarium, using water and creatures from the pond. Keeping it at home will let you study your water creatures every day, and chart their progress. Use this book to help you identify the animals you find and make a list of them.

Repeat this at regular intervals. You will find that new animals magically appear. They have hatched out of tiny eggs or changed from larvae into adults.

Most pond animals develop from eggs laid by the adults which hatch into larvae. The larvae are usually very different from the adults. Sometimes the adults live in the pond as well, for example diving beetles. Sometimes they live on land, for example dragonflies. With a bit of patience, you can find out which animals have laid eggs in your aquarium.

Make an aquarium

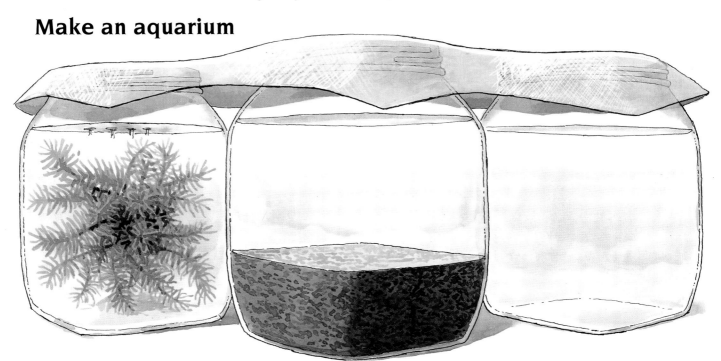

1 **Buy three cheap plastic aquaria** at your local pet store, or else use large candy jars or Tupperware boxes. If you use candy jars, they must have wide openings to allow plenty of oxygen in.
2 **Half-fill the first container with pond weed** and fill it up with pond water.
3 **In the second one, put a good layer of mud** from the bottom of the pond and fill up with pond water.
4 **Fill the third container with pond water only.** Put the containers outside in the shade and cover them with netting.
5 **Look into the containers after a day** to see what has crawled out of the weed and mud. Find out which of your containers has the widest variety of animals.
6 **Your aquaria should last for several weeks** provided they are kept cool and there is a large enough surface of water in contact with the air.

Caddis cases

If you find caddisfly cases with the larvae still in them, you can watch how they build their home. Each caddis case is open at both ends to allow a flow of water through it.

1 **If you push a matchstick very gently into the narrow end of the case,** the caddisfly larva will come out as it does not like being tickled! Now you can see it clearly. It will go back inside given the chance.

2 **Take away its old case** and keep the larva in a plastic box or jar in a cool, shady place, filled with the pond water that you found it in.
3 **Put in some tiny colored beads,** and it will rebuild its home out of these.

4 **Or you can give it pieces of the material** it normally uses, such as twigs, or shells, and watch it re-build its home.
5 **It's best to try this experiment with several caddisfly larvae,** in case one of them is lazy.

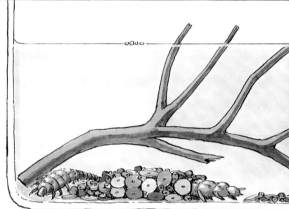

The somersaulting hydra

A hydra is a small animal that is up to about 1/2 inch long when expanded (see page 40). It looks like a hollow tube with a ring of tentacles at one end surrounding the mouth, and can move slowly by turning somersaults!

1 **Find a pond or ditch where there is a carpet of duckweed** on the surface or many floating water lily leaves.

2 **Collect water lily leaves** and look on the underside for blobs of jelly.
2 **Lift off these blobs with your paintbrush,** and leave them in water in a jar. If they are hydra, they will soon expand.
3 **Collect a quantity of duckweed,** and put it into a large jar of pond water. Keep it until the next day. Look at the sides of the jar for hydra. Watch them. They might somersault for you.

Slow Rivers & Canals

Where a river or stream reaches flat ground, its waters spread out and slow down. Mud and sand carried by the water settles out and makes a soft, muddy bottom in which many plants are able to grow. Water lilies, reeds, and sedges provide shelter for amphibians.

Hiding on the bottom and among the weeds in the warm, murky water are freshwater shrimps, mollusks, snails, and crayfish. Most lowland rivers are muddy, but rivers flowing through chalk areas are usually crystal clear. Water snails, crayfish, and shrimps thrive in these lime-rich waters, as well as a wide variety of fish. This picture shows nine animals from this book; see how many you can identify.

Mud Salamander, Mud Turtle, Chorus Frog, Mud Puppy, Iowa Darter, Swan Mussel, Crayfish, Three-spined Stickleback, Softshell Turtle.

Mud Salamander

Muddy springs and streams are the favorite habitats of this salamander, as its name suggests. The Mud Salamander is easy to spot because it is colored coral pink, bright red, or brownish salmon-pink. It has scattered black spots, and its belly is reddish or yellowish. Look in mucky areas by streamsides for this creature, but move quietly as it may burrow into the mud when it hears you coming! If you find a salamander like this in a clear stream, it may be a Red Salamander (see page 59).

Group: Lungless Salamanders (Plethodontidae)
Size: 3–8 ins
Distribution: Coastal plain and Piedmont areas, south New Jersey to central Florida

Pickerel Frog

This little, smooth-skinned frog has two parallel rows of squarish spots running down its back. The body is tan and the spots are darker. A good way to recognize it is by the bright yellow or orange marks under its hind legs, and by its whitish belly. The Pickerel Frog lives in slow-moving streams, swamps, and meadows, and has a steady, low-pitched croak. It comes out mainly at night, and gives out a nasty liquid from its skin that puts predators such as snakes off eating it.

Family: Treefrogs (Ranidae)
Size: 2–3½ ins
Distribution: Throughout eastern U.S., except Florida and extreme Southeast

Dwarf Waterdog

Group: Mud Puppies and Waterdogs (Proteidae)
Size: 4–8 ins
Distribution: Southeast U.S. along coastal plain, southeast Virginia into Georgia

This little salamander is like the Mud Puppy or Waterdog on page 12, but it is much smaller. It is dark brown or grayish black, with no patterns or markings at all. Its belly is gray, and bushy gills are visible by its neck. There are four toes on each of its feet. The Dwarf Waterdog prefers to live in slow, muddy-bottomed streams.

Reptiles

Smooth Softshell

As the name suggests, this softshell turtle has a completely smooth skin on its back, with no lumps, bumps, or spines. Its skin is olive to orange-brown, with various patterns of dots, dashes, or blotches. The females are larger than males and may have a mottled pattern on their backs. It lives in rivers and large streams with sandy or muddy bottoms. It spends most of its time in the water, feeding on crayfish, frogs, and fish. At times it basks on sandbanks, but will quickly dive back into the water if approached.

Group: Softshell Turtles (Trionychidae)
Size: Males 4½–7 ins, females 7–14 ins
Distribution: Central U.S. in the Ohio, Mississippi, Missouri, Arkansas, and Alabama drainages

Stinkpot

Other names for this little turtle are Musk Turtle and Stinking Jim. This is because it gives out a very smelly liquid if disturbed. Its shell is smooth, with a high dome, and is olive-brown in color. There are two yellow stripes on either side of its head. Its undershell is small, with lots of exposed flesh between its scales. The Stinkpot lives in slow-flowing, shallow waters that have muddy bottoms. It spends most of its time feeding on plants and decaying animals at the bottom of the water. Sometimes, it basks on overhanging trees and shrubs, but will drop back into the water if frightened. Be careful with this turtle, as males are very aggressive and can give a nasty bite.

Group: Musk and Mud Turtles (Kinosternidae) – Size: 3–5 ins
Distribution: Southeastern Canada and most of eastern U.S.
BEWARE—this animal can give you a nasty bite.

Spiny Softshell

Like all softshells, this turtle has soft, leathery skin, a long neck, and a tube-like snout. It is olive or tan in color, often with black-edged spots or dark mottling. There are spiny bumps on the front of the shell and sometimes over the back as well. The Spiny Softshell often lives in rivers, but also likes creeks, ponds, and lakes. You may find it hard to spot one—it moves fast on land and in water, where it chases minnows and other prey. Softshells also like to sunbathe on banks and logs. Be careful of this turtle, as it can give a nasty bite, and it scratches too.

Group: Softshell Turtles (Trionychidae)
Size: Males 5–9 ins long, females 6–18 ins
Distribution: Central and eastern U.S., some in Montana, southern Quebec, Delaware, and Gila-Colorado River system of New Mexico and Utah
BEWARE—this animal can give you a nasty bite.

Queen Snake

This snake is medium in size and is colored olive-brown to dark brown with a yellow stripe. Its belly is creamy-yellow. The Queen Snake lives near streams and small rivers which have rocky bottoms, and feeds almost entirely on crayfish. It often sunbathes on trees and shrubs overhanging the water, and if disturbed, it drops in. Females give birth to live young between July and September.

Group: Colubrid Snakes (Colubridae)
Size: 16–36 ins
Distribution: South Great Lakes and southeastern Pennsylvania south to Gulf Coast; some in northern Michigan, southwestern Missouri and northwestern Arkansas

Yellow Mud Turtle

The Yellow Mud Turtle gets its name from its yellow undershell, and its yellow jaw and throat. Its back shell is olive to brown. It lives in slow-moving waters with muddy or sandy bottoms, and feeds on worms, crustaceans, snails, and tadpoles. It comes out to hunt on land at dawn or dusk, and buries itself under leaves or in mud under the water when it gets cold.

Group: Musk and Mud Turtles (Kinosternidae)
Size: 3½–7 ins
Distribution: Central southern U.S. from Nebraska south to Texas and including parts of New Mexico and Arizona; some in northwestern Illinois

Mud Turtle

This turtle has an oblong, olive to dark brown, smooth shell. The undershell is yellow to brown. The head is usually dark brown with pale streaks and spots. Look for the Mud Turtle in shallow, slow-moving water where there are plenty of weeds. It spends most of its time on the muddy bottom, but often travels over land during the summer. At this time, you may see it crossing roads, where many are killed by cars. If the water dries up, the Mud Turtle can burrow into the mud and survive there until the water comes back.

Group: Musk and Mud Turtles (Kinosternidae)
Size: 3–5 ins
Distribution:
Eastern U.S.
from Long
Island to Florida
and Texas, north
in Mississippi Valley
to Illinois
and Indiana

Stonecat

This catfish is mostly brownish gray, with a white belly, chin, and barbels. Its snout hangs slightly over its lower jaw. The Stonecat lives in streams or slow-moving rivers that flow over gravel. It feeds on water insects on the bottom, as well as small fish, snails, and crustaceans. It spends most of the day hidden under a rock, and comes out at night to feed. Be careful with this kind of catfish as they have poisonous spines on the fins behind the gill cover (pectoral fins).

Group: Catfishes (Ictaluridae)
Size: Up to 12 ins
Distribution: St. Lawrence River west to Alberta; most of northern and central U.S.
BEWARE: don't touch this fish—it has POISONOUS spines.

Iowa Darter

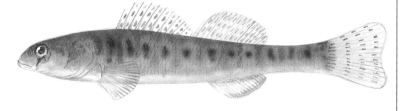

This fish has a slender body covered in medium-sized scales. An olive-brown color, it has several dark red blotches on its sides. There are two fins on its back, and the first one has eight to ten spines. The tail fin is rounded with black spots. The Iowa Darter lives in quiet areas of streams where plants grow, and in lakes with mud and sand bottoms. It feeds on insects, small crustaceans, and snails.

Group: Darters and Perches (Percidae)
Size: Up to 3 ins
Distribution: From Quebec to Alberta, New York to western Montana, south to Ohio, and west to Wyoming and Colorado

Mosquitofish

This little fish has a stout, pale-colored body covered with noticeable scales outlined in black. There are black spots on its sides, as well as on its single back fin and its rounded tail fin. It has a small head with an upturned mouth which it uses to snatch insects from the water surface. It gets the name Mosquitofish from its preferred diet of mosquito larvae. You may see this fish in swamps, ditches, ponds, lakes, and slow-moving streams. The females do not lay eggs, but give birth to live young.

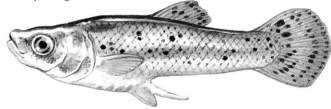

Group: Live Bearers (Poeciliidae)
Size: 2–3 ins
Distribution: Mississippi Valley from Illinois south to Texas, across Gulf Coast, northward to New Jersey and widely introduced elsewhere

Asiatic Clam

Also called the Little Basket Clam, this bivalve lives in rivers, canals, and lakes. Its shell is brown with a blackish coating on the outside, and purple or blue on the inside. It is almost triangular in shape. Look for the growth lines running along the outside of the shell. Like mussels, the clam feeds by filtering food out of the water, then ejecting the waste water, using a pair of siphons. It often takes over a stream and can be present in incredible numbers.

Group: Basket Clams (Corbiculidae) – Size: 1–2½ ins
Distribution: An introduced species now found in most of southern and central U.S—it has not yet reached Canada

Freshwater Limpet

Like the limpets you find on the seashore, freshwater limpets live firmly attached to rocks, mussel shells, stones, or plants. It can move about grazing on vegetation, but will clamp down firmly if threatened. The shell is cone-shaped, with a curved-back top. Some of these limpets like the still water of ponds and lakes. Others prefer streams and rivers. River-dwelling limpets have tougher shells to withstand the flow of water.

Group: Freshwater Limpets (Ancylidae)
Size: Up to ½ ins
Distribution: Found across North America

River Snail

This shell of the River Snail, or Nerite, is thick and patterned, with a big, D-shaped opening.It may have zigzag lines or be plain in colour. It lives mostly in running water such as streams and rivers, but may also be seen in canals and ponds. When the snail retreats into its shell, it seals the opening with a lid or "operculum."

Group: Nerites (Neritidae) – Size: Up to ¾ in
Distribution: Found across North America

Swan Mussel

Many sorts of mussel live in fresh water and it can be difficult to tell them apart. The Swan Mussel is one of the largest. It has an oval shell flattened on one side. Yellowish green to olive-brown is their normal color, but they are often stained a dirty black. In contrast, the inside of the shell is a lovely, iridescent mother-of-pearl. It lives in ponds, canals, and slow-flowing rivers. Food is filtered from the water as it lies half-buried in the mud.

Group: Freshwater Mussels (Unionidae)
Size: Up to 6 ins
Distribution: Found across North America

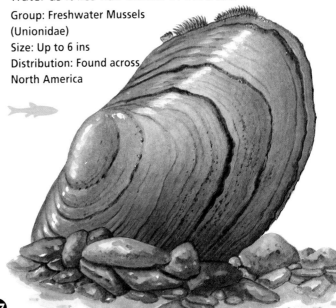

Find Out Some More

Useful Organizations

In addition to the national groups listed below, most states have their own herpetological society, and there are hundreds of local natural history associations. Check with your teacher, or with your nearest natural history museum, wildlife refuge, or local public library for information on them.

North American Native Fishes Association is interested in studying and conserving American fish. Write to: North American Native Fishes Association, c/o Bruce Gebhardt, 123 West Mt. Airy Avenue, Philadelphia, Pennsylvania 19119.

Society for the Study of Amphibians and Reptiles is a good group for you to contact. Write to: Society for the Study of Amphibians and Reptiles, c/o Dr. Douglas Taylor, Miami University, Dept. of Zoology, Oxford, Ohio 45056.

The **American Society of Ichthyology and Herpetology** is a group for professional zoologists and serious amateurs (ichthyology is the study of fish, herpetology is the study of amphibians and reptiles.) Write to: American Society of Ichthyology and Herpetology, Department of Zoology, Business Office, Southern Illinois University, Carbondale, Illinois 62901–6501.

Reptile and Amphibian Magazine is published every second month. To subscribe to it, write to: Reptile and Amphibian Magazine, RD#3, Box 3709–A, Pottsville, Pennsylvania 17901.

Many of the preserves owned by the **Nature Conservancy** and its chapters conserve unique and threatened habitats for amphibians and reptiles. Write to: Nature Conservancy, Suite 800, 1800 North Kent Street, Arlington, Virginia 22209.

In Canada, the **Canadian Nature Foundation** is a good starting point. Write to: Canadian Nature Foundation, 453 Sussex Drive, Ottawa, Ontario K1N 6Z4.

Places To Visit

Lake Ontario, New York. The best time to see the fish is during the spawning season. Visit the state fish hatchery at Altmar. The Niagara River at Lewistown has huge numbers of smelt in April; in spring and fall, the Salmon River near Pulaski fills with intrduced Pacific salmon and steelhead.

Back Bay National Wildlife Refuge, Virginia Beach, Virginia. Tidal marshes, freshwater ponds, and brackish (slightly salty) ponds are homes for snapping turtles, mud turtles, and other species.

Great Smoky Mountains National Park, Tennessee–North Carolina. This is the home of 28 different species of salamander—North America's largest collection. The moist forests shelter over 120 species of amphibians and reptiles.

Everglades National Park, Flamingo, South Florida. This "river of grass" is the home of water snakes, American alligators, cottonmouths, softshell turtles, plus many other amphibians.

Corkscrew Swamp Sanctuary, Naples, Florida. Owned by the National Audubon Society, there is a two-mile-long boardwalk through the cypress swamps where you may see tree lizards, American alligators, and a wide variety of snakes.

Buffalo National River, Harrison, Arkansas. Flowing for 132 miles through the Ozarks, it holds smallmouth bass, various sunfish and darter species, and other cool-water species.

Big Bend National Park, Texas. The Rio Grande River provides a home for many amphibians including the leopard frog.

Le Hardy Rapids, Yellowstone National Park, Wyoming. Every June, these violent rapids on the Yellowstone River fill with cutthroat trout fighting their way up to spawn in Yellowstone Lake.

There are also many preserves and sanctuaries in every state—check with your teacher, your nearest natural history museum, wildlife refuge, or local public library for information on them.

Index & Glossary

Useful Books

Animals of Ponds and Streams, Julie Becker (EMC Publishers).

Audubon Society Field Guide to North American Reptiles and Amphibians, John L. Behler & F. Wayne King (Alfred A. Knopf). A photographic guide that describes all the species found in North America north of Mexico.

Encyclopedia of Aquatic Life, Dr. Keith Banister & Dr. Andrew Campbell (Facts on File). A comprehensive guide to plants and animals.

Peterson First Guide to Reptiles and Amphibians (Houghton Mifflin Co.). A simplified pocket-sized guide to common species.

Peterson Field Guide to Freshwater Fishes, Lawrence M. Page (Houghton Mifflin Co.). A pocket-sized guide to common species.

Thw World of Dragonflies and Damselflies, Roos E. Hutchins (Dodd, Mead Co.).

Index & Glossary

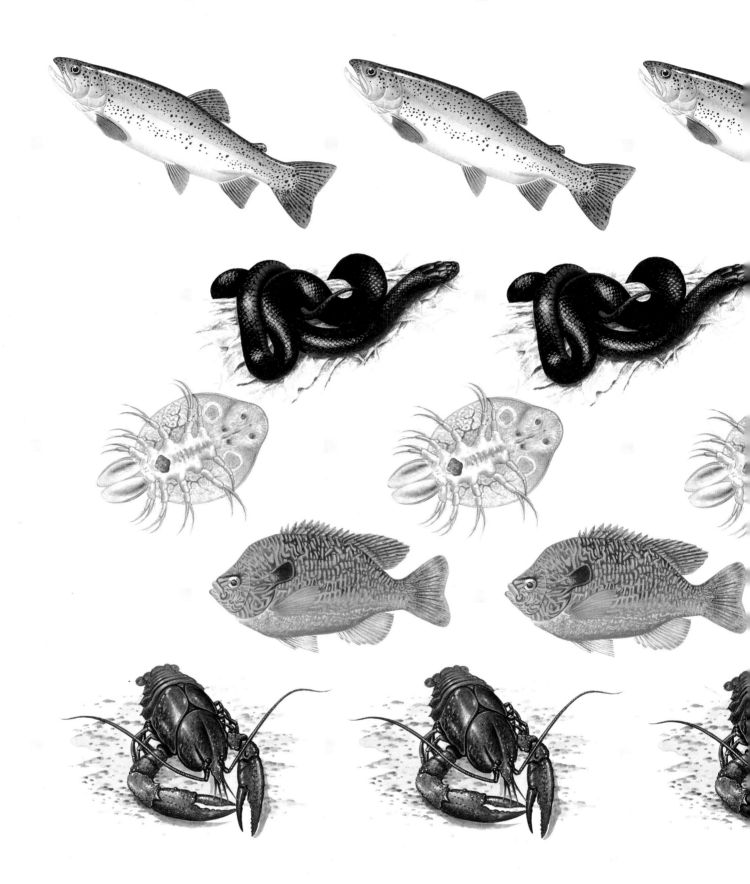